Sinja Hantscher



# The UNHCR and Disaster Displacement in the 21st Century

## An Organizational Analysis

Springer

Sinja Hantscher
Department of History and Social Sciences
Institute of Political Science
TU Darmstadt
Darmstadt, Germany

ISSN 2198-7289 ISSN 2198-7297 (electronic)
Contributions to Political Science
Dissertation 2018 Technische Universität Darmstadt
ISBN 978-3-030-19688-2 ISBN 978-3-030-19689-9 (eBook)
https://doi.org/10.1007/978-3-030-19689-9



This Springer imprint is published by the registered company Springer Nature Switzerland AG.
The registered company address is: Gewerbestrasse 11, 6330 Cham, Switzerland

*To my family*

# Acknowledgments

First of all, I would like to thank Prof. Dr. Markus Lederer and express my gratitude for his invaluable support, dedicated supervision, and continuous motivation throughout the last few years. I consider myself very lucky to have been in his research group—thank you for making me a part of *Team Lederer*! Furthermore, I would like to thank Prof. Dr. Klaus Dingwerth. I am deeply grateful for his input on the research design of this project. His expertise and remarks contributed immensely to this thesis' theoretical approach. In addition, I am grateful to Prof. Dr. Jens Steffek and Prof. Dr. Ulrich Brinkmann for agreeing to evaluate my dissertation.

I would also like to thank Judith Kreuter, Chris Höhne, Linda Wallbott, and all the other great colleagues of the *Team Lederer*-retreats—your feedback and support during the past few years have really meant a lot to me! I am also grateful for the support I received in data collection by Bianca Creutz, Sophie Hölscher, Henrik Schaper, Magdalena Tröndle, and Tim Huß. Adding to this, I am very thankful for having been part of the Graduate School of Politics at the University of Münster and for input from the colleagues at the Technical University of Darmstadt and the University of St. Gallen. Special thanks go to Christiane Bomert for being such a great library companion and for always having my back; to Anne Hennings for her "you've got this" mentality and her support; and to Dr. Stephanie Tremmel and the colleagues at the German Federal Police for their motivation during the last few months! In addition, I owe thanks to my interview partners in Geneva for their time, their interest, and their openness to discuss the topic of EDPs—this thesis benefited hugely from your valuable insights! Thanks also to the editors and the team at Springer Nature who were all very helpful and efficient at getting the book through the production process.

Last but not least, I am deeply thankful to my friends for the unconditional support during the last few years. I would like to express my deep gratitude to Alexander Roesler and my family for always believing in me and for all the encouragement! Thank you!

# Contents

# List of Abbreviations

| | |
|---|---|
| AR | Academic representative (transcript of expert interview) |
| COP | Conference of Parties |
| ECOSOC | Economic and Social Council |
| EDP | Environmentally displaced person |
| e.g. | for example |
| ERC | Emergency Relief Coordinator (OCHA) |
| ExCom | UNHCR Executive Committee |
| GA | UN General Assembly |
| IASC | Inter-Agency Standing Committee |
| IDMC | Internal Displacement Monitoring Centre |
| IDP | Internally displaced person |
| i.e. | that said |
| IO | International organization |
| IOM | International Organization for Migration |
| IOR | IO representative (transcript of expert interview) |
| IPCC | Intergovernmental Panel on Climate Change |
| IR | International Relations (= field of political science; international relations = global affairs/politics) |
| IRO | International Refugee Organization |
| MECLEP | Migration, Environment and Climate Change: Evidence for Policy Project |
| MNCs | Multinational corporations |
| NAPs | National Adaptation Plans |
| NGO | Non-governmental organization |
| NGOR | NGO representative (transcript of expert interview) |
| NRC | Norwegian Refugee Council |
| OCHA | United Nations Office for the Coordination of Humanitarian Affairs |
| OHCHR | Office of the High Commissioner for Human Rights |
| p. | Page |

| | |
|---|---|
| para. | Paragraph |
| Refugee Convention | United Nations Convention Relating to the Status of Refugees |
| SR | State representative (transcript of expert interviews) |
| UN | United Nations |
| UNEP | United Nations Environmental Programme |
| UNESCO | United Nations Educational, Scientific and Cultural Organization |
| UNFCCC | United Nations Framework Convention on Climate Change |
| UNFPA | UN Population Fund |
| UNHCR | United Nations High Commissioner for Refugees |
| UNU | United Nations University |
| USG/ERC | Under-Secretary-General and Emergency Relief Coordinator of OCHA |

# List of Figures

# List of Tables

# Chapter 1
# Introduction: Disaster Displacement

> *It cannot be contested that climate change will be a significant influence on migration, both internal and international, in the later decades of the twenty-first century.*
>
> (Hugo 2010, p. 9)

Climate change, hazards, and environmental changes in general constitute just one trigger for migration and displacement among other social, economic, political, and other factors. However, the prospect of a potential future in which whole stretches of land would become uninhabitable and where millions of people would lose their livelihoods due to environmental factors is especially daunting, because the culprits of migration and displacement are not as black-and-white and are harder to identify than in a war and conflict scenario. Even though this thesis will show that forecasting in the context of environmental changes and the possible human mobility that comes along with it is particularly challenging, this has not prevented scaremongering and the circulation of grotesque future scenarios during the last decades. In this context, migration expert Graeme Hugo declares that environmental changes have an undeniable impact on migration and displacement—and this environmental impact on human mobility will grow even larger in the future. In a similar vein, one of the leading legal scholars on the topic, Walter Kälin, argues that there exists a 'normative gap' with regard to people displaced across borders due to sudden-onset disasters, by which he means that these people do not fall under any legal and binding protection framework (Kälin 2010, p. 89). Similarly, those people who are being displaced because of slow-onset environmental changes have until recently only received limited political attention. According to Kälin, "The status of these people [environmentally displaced persons] remains unclear, and despite the applicability of human rights law, there is a risk that they may end up in a legal and operational limbo" (Kälin 2010, p. 89). This thesis uses the term 'environmentally displaced persons' (EDPs) to refer to "those forcibly displaced at least partly because of a natural disaster—whether it is climate-related or not" (Kolmannskog 2012, p. 1072).

The last decade has shown a rise in attention toward EDPs at international forums and in national politics, and a wide range of actors have picked up the topic,



S. Hantscher, *The UNHCR and Disaster Displacement in the 21st Century*, Contributions to Political Science, https://doi.org/10.1007/978-3-030-19689-9_1

including states, international organizations (IOs), and non-governmental organizations (NGOs). Particularly with regard to environmental migration and displacement, IOs are important actors. The variety of actors include organizations such as the International Organization for Migration (IOM), the United Nations High Commissioner for Refugees (UNHCR), and the International Labor Organization (ILO), as well as the Office of the High Commissioner for Human Rights (OHCHR), the United Nations Population Fund (UNFPA), and the World Bank (Betts 2011, p. 16).

The thesis at hand takes the above-mentioned protection gap of EDPs and the recent rise of the topic in international forums as starting points. The research interest of this thesis concerns the role of IOs in the rise of the topic as well as the underlying goals that IOs might attach to picking up the topic of EDPs. The thesis thereby intends to contribute to the research on IOs and to the research on EDPs. For this reason, the following sections will, firstly, introduce the topic of disaster displacement; secondly, outline the empirical puzzle that this thesis will focus on; and thirdly, illustrate the approach of the research project. Lastly, this introductory chapter will discuss the empirical and theoretical contributions of this thesis and will present the structure of the following chapters.

## 1.1 Disaster Displacement

The topic of environmental migration and displacement arose on the agenda around the mid- to late-1990s, and especially before the negotiations of the Kyoto Protocol (NGOR2, para. 63). NGOs advocated the topic—especially due to climate change—by means of a scare tactic, and these organizations introduced lurid headlines of waves of 'climate refugees'. By this means, NGOs intended to sensitize the global North to the future impacts of environmental changes in their countries. The underlying intention was to push states in their efforts to strengthen the mitigation of climate change (NGOR2, para. 63). A similar depiction shows itself in the academic literature, one example being the publication by Norman Myers and his declaration about the future number of 'climate refugees' (Myers 1993).

In the 2000s, the debate slowly began to shift toward a more nuanced perspective on the topic. The number of publications increased, and the tenor changed into a new depiction in which the impacts of environmental changes become manageable if the international community puts more effort into adaptation (NGOR2, para. 63).[1] Hereby, the intention was to provide evidence about likely future developments and to develop tools to adapt to future changes (NGOR2, para. 63). Yet, the rhetoric in the media on environmental migration and displacement largely evoked an image of uncontrollable and alarming future patterns of human mobility. By referring to

[1]In the context of EDPs, adaptation means "designing and implementing measures to help communities affected [...] to modify the ways they work and live to be able to cope with new environmental conditions" (Castles 2010, p. 240).

EDPs as refugees, the term tends to convey the image of a person who is possibly threatening and illegal (Elliott 2010, p. 183). Until around 2010, references to national or European security concerns were very common and depictions in the context of EDPs often called for an expansion of military capacities (Elliott 2010, pp. 183–184).

Despite this scaremongering, the image of EDPs as "a threat to the security, prosperity and public health of rich countries in the global North" (Castles 2010, p. 242) contradicts the scientifically grounded predictions. Disaster displacement research predicts that many EDPs will remain within their country of origin and thereby fall under the legal framework of the Guiding Principles of Internal Displacement, which draws on binding principles of refugee law, international humanitarian law, and human rights law (Kolmannskog 2012, p. 1072). The Guiding Principles themselves set out principles for protection but are non-binding (Zetter 2010, p. 133). Additionally, the extent of migration and displacement after extreme events depends on whether the country of concern has a history of migration. Human mobility[2] can be explained by already existing "social or cultural territory on which a community is based" (Findlay 2011, p. 53) which makes international or long-distance migration and displacement unlikely. Additionally, most EDPs return to their local areas and will participate in post-disaster recovery (among others, Barnett and Webber 2010, p. 40; IPCC 2012, p. 293). An exception to this is cases in which environmental changes destroy the basis for living in a place—here, displacement might become permanent (IPCC 2012, p. 293) What is more certain is that both the impacts of environmental changes and the consequences of migration or displacement are human security issues—that is, there is a focus on adaptation and building resilience (among others, Elliott 2010, p. 190).

The spheres of governance that apply in the context of EDPs are diverse and involve a variety of responsible IOs. McAdam differentiates between migration/asylum, the environment, development, disaster response, and human rights as fields of governance, and she lists the UNHCR, the IOM, the OHCHR Global Migration Group, ILO-MIGRANT, the UNFPA, and the Internal Displacement Monitoring Centre (IDMC) as IOs which involve themselves in the topic (McAdam 2011), pp. 168–169). More specifically, various global actors monitor access to resources and mediate vulnerability in the case of environmental catastrophes, very often down to the local level (Mearns and Norton 2010, p. 65).

### *1.1.1 The State of the Art on Disaster Displacement*

Building upon the above-mentioned perceptions of EDPs and the diverse range of international actors in this policy field, the focus will turn to how academia has picked up the topic of disaster displacement. The non-academic literature has mostly

---

[2]People are only able to overcome the financial, information and legal barriers of migration if members of their family or community have already established a support network at the new location (Barnett and Webber 2010, p. 42).

linked environmental changes and human mobility in dramatic terms. The main debate includes millions of 'environmental refugees'—such as 50 million by 2020 or even up to 250 million people by 2050—despite the lack of coherent evidence for these statements (Black et al. 2011b, p. 3; Subramanian and Urpelainen 2014, p. 26). Contrary to this view, the current research on the link between environmental changes and human mobility paints a rather different picture. It is difficult to establish a simple relationship between environmental changes and human mobility (Collins 2013, p. 113). Overall, the environment is seen as *one* driver of human mobility, regularly occurring in combination with other socioeconomic drivers. Following this line of thought, most authors have stressed that it is quite rare specifically to identify migrants who are purely driven by environmental concerns (Black et al. 2011b, p. 4). Responding to the fear of mass migration of some countries from the global North, the research has shown that the developing countries will receive the largest amount of permanent migration due to environmental changes (Barnett and Webber 2010, p. 51), even though they have the lowest receiver capability.

As regards the argumentation of so-called 'mass migration' due to (slow- or sudden-onset) environmental changes, researchers have been careful with these kinds of assertion because the data are weak. Specifically, studies[3] on mass migration associated with coastal floods due to sudden-impact disasters have shown that there are alternatives to migration, thus questioning the occurrence of huge numbers of migrants due to these kinds of disaster (Black et al. 2011b, p. 4). Generally speaking, the empirical evidence for 'mass migration' due to environmental changes is very weak. Some authors, such as Portes and de Wind (Portes and Wind 2007), have even gone so far as to question, first, the idea of drivers of migration in general, and second, the possibility of developing a comprehensive theory of migration in particular. Furthermore, when asked, the displaced usually stress socioeconomic factors, even when specifically quizzed about environmental factors (Black et al. 2011b, p. 5).

The literature shows that the topic of environmental migration and displacement has been repeatedly discussed in the context of security—e.g., the relationship between security and environmental changes, or the topic of EDPs from a security perspective (among others, Baldwin and Bettini 2017; Boas and Rothe 2016; Corry 2012; Kita and Raleigh 2018; von Lucke et al. 2014; Meierding 2013; Methmann et al. 2013; Oels and von Lucke 2015; Rothe 2012, 2016; Saha 2012; Scheffran et al. 2012; Tänzler and Ries 2012; Trombetta 2014). In addition, the global environmental governance regime has received considerable attention (among others, Biermann et al. 2009a). In this context, it is not only the natural science perspective—e.g., the specifics of climate change—that has received scholarly attention, but also the global actors who involve themselves in the topic of EDPs. That said, scholars have analyzed the role of states, on the one hand, and IOs, NGOs, businesses, and the civil society on the other hand in the global (environmental) governance regime (Goodwin-Gill and McAdam 2017; McAdam 2017; McLeman and Gemenne

[3]For more information, see Black et al. (2011a) and Warner et al. (2010).

2018b; Steffek and Kortendiek 2018). Among this variety of actors, IOs and their respective involvement in EDP matters have recently come to the fore research-wise. First, scholars acknowledge that IOs take part in global environmental negotiations, influence the conclusion of contracts, and are important bearers of knowledge (Franck 2018; Hall 2016a; Martin et al. 2018). Second, the way that an IO's involvement in the global environmental regime reflects internally—i.e., in the IO's decision-making, bureaucratic processes, and organizational direction—has also received some academic attention (Goodwin-Gill and McAdam 2017; Hall 2016b; Rosenow-Williams and Gemenne 2016; Sosa-Nunez and Atkins 2016). Therefore, the next section will focus on how one IO, the UNHCR, has approached the topic of EDPs so far.

### 1.1.2 The UNHCR and Disaster Displacement

The UNHCR is the leading global refugee agency with a focus on rights-based and protection concerns. Furthermore, the organization stands out among the above-mentioned international actors because its interest in the topic of EDPs is particularly puzzling. The UNHCR is an IO that is mostly concerned with refugees and has rather little historical experience concerning environmental changes; yet, the organization shows tendencies to extend its work to people displaced due to environmental changes—whether the environmentally displaced have to relocate in their home country or to a foreign country (see Sect. 5.2).

Concerning its approach to EDPs, the UNHCR argued that the terms 'voluntary migration' and 'displacement' need further elaboration when applied to natural disasters. Especially since the early 2000s, the UNHCR has voiced concern for those displaced due to environmental changes. As of 2005, there exists a tendency at the UNHCR to involve itself in protection activities. The UNHCR acknowledged that it has "an interest in such movements of people and an ability to respond to their needs" (UNHCR 2012, p. 26). Former High Commissioner António Guterres approached the issue of EDP protection at an Executive Committee (ExCom) meeting in 2007. According to Guterres, climate change will force a rising number of people to move due to environmental degradation. Hence, the UNHCR's role will be to "alert states to these problems and help find answers to the new challenges they represent" (Guterres 2007, pp. 1–2). Due to Guterres' lead on the issue of environmental displacement, the UNHCR published its first policy paper on the topic in 2008 (UNHCR 2008), and the organization further participated in the UN's Inter-Agency Standing Committee (IASC) that dealt with the issue (McAdam 2011, p. 8). The accounts of the UNHCR in this context were remarkable: The official publications, the displays on their website, and the tones of their official spokesperson and executive head largely depicted a grim future for millions of people who will become EDPs and who will cross national borders for refuge and in desperation for a better life.

In the 2010s, the UNHCR pleaded for greater nuance when applying existing legal frameworks—e.g. the Refugee Convention—to the issue of environmental changes. In 2012, the organization stated that it would be "prepared to work with states and other actors to develop a guiding framework or instrument to apply to situations of external displacement outside those covered by the 1951 Convention, and in particular to displacement resulting from climate change and natural disasters" (UNHCR 2012, p. 28). The UNHCR thereby cooperated extensively with other international actors (see Sect. 5.2.1) and has worked together with other IOs on the Operational Guidelines on Human Rights and Natural Disasters. The IASC adopted these guidelines in 2006 to guarantee the human rights of those people who suffer under natural disasters, whether or not they are displaced (Kälin 2010, p. 93). The cooperation between the UNHCR and NGOs has extended throughout the existence of the organization and includes NGOs as implementing partners as well as NGOs sitting on the ExCom (Dingwerth et al. 2014, p. 198). When looking at the UNHCR's approach to EDPs as depicted above, it shows that the organization's behavior represents an empirical puzzle, which the subsequent paragraphs will expand on.

## 1.2 The Empirical Puzzle and the Research Gap

The UNHCR is not the only IO that involves itself in the topic of EDPs.[4] However, the UNHCR stands out because the topic of EDPs is in stark contrast to its official mandate and workload. According to Article 1 of the Refugee Convention, the UNHCR's mandate only applies to refugees—i.e., to people who flee across a national border because of persecution due to race, religion, nationality, membership of a particular social group, or political opinion (UNHCR 1951). In other words, EDPs do not fall under the mandate of the UNHCR, which means that its mandate does not cover the protection of these people. Therefore, the role of the UNHCR on questions of environmentally-induced migration or displacement is controversial: the organization is already responsible for over 20 million refugees and it is financially dependent on states for funding (UNHCR 2017). Despite this, the UNHCR further involves itself in the topic and is rather active in the global arena when it comes to EDPs (see Sect. 5.2). That said, the reason for choosing the UNHCR as the object of investigation is the underlying interest in the ambivalence of the organization.

From an academic perspective, research is still in its fledgling stages concerning the question of how the leading migration and refugee actors are dealing with the topic of EDPs (cf. Chazelnoel and Ionesco 2018; Franck 2018; Hall 2016a; McAdam 2017). IOs are the actors who largely address the topic of environmental migration and displacement, and that is why scholars need to pay more attention to the way these global players move in this field and why they do so. This thesis aims to add to the existing research on this topic. As an object of investigation, the UNHCR has received increasing scholarly attention since the 1990s. Volker Türk, who later

[4]Compare the IOM and the Norwegian Refugee Council (NRC) (Chazelnoel and Ionesco 2018).

became the Assistant High Commissioner for Protection, published on the history and the mandate of the UNHCR (Türk 1992). Added to this, Barnett and Finnemore (Barnett and Finnemore 1999) incorporate the UNHCR in their work on IO pathologies as one of the case studies. In this context, they use the UNHCR as an example to point out deviations from the mandate and outline how this deviant behavior becomes normal over time. Additionally, Barnett and Finnemore outline the cultural battle inside the UNHCR between the headquarters and the field offices—e.g., on the topic of repatriation (Barnett and Finnemore 1999, p. 725). Dingwerth et al. (2014) focus on the UNHCR and its addressees—in other words, the actors who need to authorize and approve the UNHCR's decisions and policies—and Hall (2015) approaches the UNHCR's position on climate change by means of an inter-organizational comparison. By focusing on climate change, Hall analyzes the UNHCR approach in comparison to the approaches of the IOM and the World Health Organization (WHO), carving out the specific characteristics of the respective organization (Hall 2015, p. 79). Lastly, scholars have already done preliminary work on the way the UNHCR interprets its mandate. Barnett and Finnemore declare that the UNHCR has grown in terms of autonomy and authority (Barnett and Finnemore 1999, p. 705). They trace this organizational growth back to the UNHCR's long years of experience concerning refugee protection and its official standing in the refugee regime which led to an acknowledgement of its expert status (Barnett and Finnemore 1999, p. 710).[5]

This thesis builds upon this previous research, but it will also tackle the research gap that exists regarding the question of why the UNHCR approaches the topic of EDPs the way that it does. The recent research by Nina Hall is the first step toward shedding light on this gap. She chooses the UNHCR as one of her case studies and asks whether the organization moves beyond its mandate to respond to climate change (Hall 2016a). Hall thereby conducts a similar organizational analysis of the UNHCR, but with an exclusive focus on climate change. The thesis at hand takes Hall's hypotheses into consideration but goes one step further. Whereas Hall centers her analysis on the interest of various IOs in climate change, this thesis constitutes an in-depth case study of only one IO—the UNHCR. Added to this, this thesis exceeds Hall's limit of only considering climate change; instead, in this thesis, the UNHCR's interest in EDPs takes center stage, irrespective of the environmental change, be it climate change or not, that led to the displacement. The underlying reason for this broader research approach is that it avoids jumping on the bandwagon of the buzzwords 'climate change'. Therefore, the focus on EDPs provides an insight into the research gap that exists on the UNHCR's interest in matters concerning environmental changes generally. Following these observations, the question arises:

> Why does the UNHCR approach the issue of EDPs the way that it does?

---

[5]For an overview of the research on the UNHCR, see Sects. 3.3.2 and 5.1 of this thesis, or Goodwin-Gill and McAdam (2017); Hall et al. (2015); Hammerstad (2014); Loescher and Milner (2011).

Before going in-depth on the research design of this thesis, this section introduces important keywords: migration, displacement, refugee, and international organization. To begin with, the question of EDPs is part of the broader context of international migration. In other words, the empirical puzzle of this thesis falls into the broader debate on how to approach international migration and human mobility in general. Until the IOM became part of the UN family in 2016, the question of migration was not properly tackled in the international forums; rather, every state approached the topic of migration in an individual manner, irrespective of cross-border or internal migration. For the purpose of this thesis, and in line with the terminology suggested by the Nansen Initiative and the Cancun Climate Change Adaptation Framework, migration refers to "human movements that are preponderantly voluntary insofar as people, while not necessarily having the ability to decide in complete freedom, still possess the ability to choose between different realistic options" (The Nansen Initiative 2015, p. 29).

The second important keyword is 'displacement'. In this context, displacement is understood as the "primarily forced movement of persons" (The Nansen Initiative 2015, p. 29). More specifically, people can be displaced inside a country and thus become internally displaced persons (IDPs). In contrast, cross-border displacement falls—under certain conditions—under the category of the Refugee Convention, and the UNHCR is responsible for these persons. In the context of EDPs, there are differences between sudden- and slow-onset environmental triggers: whereas the first trigger often leads to short-term displacement, the latter might lead to long-term displacement. In both cases, cross-border displacement (short-term) or migration (long-term) is a possibility. For this analysis, disaster displacement "refers to situations where people are forced to leave their homes or places of habitual residence as a result of a disaster or in order to avoid the impact of an immediate and foreseeable natural hazard" (Platform on Disaster Displacement 2018). In this context, sudden-onset disasters are both meteorological hazards (e.g., flooding, windstorms, or mudslides) and geophysical hazards (e.g., earthquakes, tsunamis, or volcano eruptions), whereas slow-onset disasters comprise environmental degradation processes (droughts, desertification, salinization, or rising sea levels) (Platform on Disaster Displacement 2018).

'Refugee' is the third keyword, and the Convention Relating to the Status of Refugees defines the term in its Article 1 proclamation. According to the Refugee Convention, a refugee is "any person who [...] owing to well-founded fear of being persecuted for reasons of race, religion, nationality, membership of a particular social group, or political opinion, is outside the country of his nationality and is unable or, owing to such fear, is unwilling to avail himself of the protection of that country" (UNHCR 1951). Contrary to this definition, those people who move across a border due to environmental changes such as natural disasters or climate change do not fall under any legally binding concept that protects them. Thereby, EDPs depend upon the goodwill of the neighboring countries or the international community. A *migrant*, a person who is *displaced*, and a *refugee* are therefore three different categories of human mobility and the term 'EDP' refers to "those forcibly displaced at least partly because of a natural disaster—whether it is climate-related or not" (Kolmannskog 2012, p. 1072).

As mentioned above, the UNHCR—and thus an IO—constitutes the case study for this thesis. Following Biermann and Siebenhüner, an IO is an "institutional arrangement that combines bureaucracies with a normative framework that is set by and is effective on states (and sometimes on non-state actors)" (Biermann and Siebenhüner 2009, p. 7). Added to this, Barnett and Finnemore point out that IOs are bureaucracies (Barnett and Finnemore 2004, p. 3). This thesis concurs with Barnett and Finnemore and thus uses the terms 'IOs' and 'international bureaucracy' interchangeably.[6] For the aim of this thesis, an IO is a purposeful and (mostly) autonomous and legitimacy-driven actor that moves in the international arena and has a definable membership and a hierarchical structure. As regards terminology, this thesis uses the term 'organizational research' for research on organizational theories and the term 'research on IOs' for research on international organizations.

## 1.3 The Approach of the Research Project

This section outlines the design of this research project, first by explaining the case selection, then introducing the theoretical approach of the project, and, third, stating the methodology.

### *1.3.1 The Case Selection*

The case study of this thesis is chosen based on careful consideration of the following aspects. Even though a lot of actors play an important role in the governance of refugees, the UNHCR still represents the center of the refugee regime and the Refugee Convention is "the centerpiece of international refugee protection today" (UNHCR 2010, p. 2). The topic of refugees gained momentum after the Second World War against the background of the extreme human rights violations of the Holocaust (Loescher and Milner 2011, p. 190). The Universal Declaration of Human Rights of 1948 is the basis for the 1951 Convention, because Article 14 of the Declaration acknowledges the right to seek asylum from persecution (UNHCR 2010, p. 2). As the central research question already indicates, the aim of this study is not to compare different IOs concerning their respective approaches toward environmental migration and displacement. Rather, the intention is to focus on one IO, the UNHCR, because the organization seems to behave inconsistently with its mandate. Therefore, the empirical puzzle directly concerns the UNHCR and its

[6]With respect to institutions and regimes, this thesis sees these as "abstract sets of principles, norms, rules, and procedures that do not possess a material entity of their own" (Biermann et al. 2009b, p. 39).

approach to EDPs. In other words, the case of this thesis stands out in comparison to other possible cases—other IOs—and this explains why this thesis is a single case study analysis. In the existing research on case selection, the approach of this thesis is accepted as one way among many others to tackle a research problem: that is to say, in a single case study "the problem of case selection does not exist (or is at any rate minimized), for the case of primary concern has been identified a priori" (Seawright and Gerring 2008, p. 296).

Added to this, it is apparent that there exists a lack of representativeness in single case studies (George and Bennett 2005, p. 30); single case studies do not even intend to be representative (Leuffen 2007, p. 204). The findings of this analysis are not sufficient to apply to other cases—i.e., other IOs. It is scientifically incorrect to make direct claims from one case study to another. That is not to say that a single case study does not ultimately lead to empirically and theoretically relevant results. Instead, it depends on the overall aim of the project and the empirical or theoretical puzzle at hand whether a single case study or a comparative analysis makes sense. Single case studies can be useful in those cases when a scientist wants to test a theory or wants to develop a theory by means of multiple observations of one single case (George and Bennett 2005, p. 32). Therefore, single case studies are used best in those approaches that intend to follow a theory-led process (Leuffen 2007, p. 204). Furthermore, single cases try to understand what is peculiar about the specific case, and they go into more depth as compared to research designs that compare different cases and consequently have a more comparative design (Krumm 2009, p. 100). In the context of this thesis, a single case study is advisable because the research project is an in-depth analysis and all of the above-mentioned arguments apply. Having decided on the case, the question arises as to how the thesis will tackle the empirical puzzle theoretically. The next section therefore introduces the theoretical approach of this thesis.

### 1.3.2 The Theoretical Approach

As mentioned above, multiple international actors interact on the matter of EDPs in the international realm. Added to this, the international nature of the empirical puzzle shows itself because environmental changes do not occur in accordance with national borders. Instead, the displacement and migration due to environmental changes is a topic of global governance which both affects and is affected by international politics. Even though the obvious step would be to turn to International Relations (IR) theory, it is quickly seen that IOs are not the primary objects of analysis in IR [also see Sect. 3.1; Enkler and Grohs (2017, p. 255)]. As summed up by Koremos and colleagues, "in broad IR theories, institutions play only a modest role" (Koremenos et al. 2001, p. 762). Nevertheless, IR scholarship has undergone a paradigm shift that has led to organizational theories being encompassed within the study of global affairs (Fioretos 2011, pp. 367–368). This 'institutional turn' introduced organizational theories from both a rationalist and a constructivist background into the field of IR.

With reference to the previously depicted summing up of organizational research—by among others (Bauer et al. 2009; Dingwerth et al. 2009, p. 16;

Kratochwil and Ruggie 1986, p. 775; Martin and Simmons 2013; Verbeek 1998, pp. 11–16)—the process of how interest in IOs emerged can be elucidated through several phases. These phases coincide with the impulses of IO foundations throughout the centuries (for a detailed historical account, see Rittberger and Zangl 2003, pp. 50–87). The first phase of scholarly interest in IOs goes back to the 1950s and 1960s, but the number of publications on IOs then declined. The reason that IOs were of secondary importance in the 1970s is that the predominant worldviews of idealists and realists at that time focused only on their respective interpretations of harmony and conflict; the conditions under which states cooperate and questions concerning conflict or dependency had not yet come to the fore (Kleine 2013, pp. 324–325). The second phase, which began in the 1990s and is ongoing, is characterized by a rising interest in the UN family and in regional organizations (Dingwerth et al. 2009, pp. 13–14). According to Nölke, interest in organizational theory originated in the discovery that transnational networks increasingly influence political decision-making (Nölke 2000, pp. 335–336)[7] Dingwerth et al. (2009) trace the rising interest in IOs back to the end of the Cold War and the renewed demand for global governance, a demand that IOs covered. In a similar vein, interest in IOs has a long tradition in the research on political parties, public policy analysis, and administrations ('Verwaltungsforschung') (Dingwerth et al. 2009, pp. 14–15)[8] Overall, the intention of organizational research is to develop theories about organizations as diverse as firms, schools, and bureaucracies (Dingwerth et al. 2009, pp. 18–19).

Furthermore, the perspectives on IOs from the principal-agent and bureaucracy approaches focus on two separate but complementary aspects of IOs. The former approach helps to tackle questions that relate to the external and organization–member state relations of IOs, whereas the latter sheds light on the internal dynamics, such as organizational culture (Dingwerth et al. 2009, p. 18; Hanrieder 2014b, p. 225). To put it another way, Ege and Bauer (2013) call the first approach the IR approach, which answers the question of IO creation, whereas the second approach is a public administration view of IOs, which approaches questions of policy-making in bureaucracies (Ege and Bauer 2013, p. 135). Concerning their interpretation of the state of the art of research, Ege and Bauer fit into the above-mentioned rough dichotomy of the principal-agent approaches with their rational-choice institutionalism background and constructivist approach, along with their history of sociological institutionalism (Ege and Bauer 2013, pp. 138–139).

Therefore, the current state of research has benefited from input from international relations, comparative political research, and administrative sciences in a complementary fashion. This enables political scientists to get a glimpse of inner-organizational processes, intra-organizational relations, and the influences of IOs in the international sphere (Dingwerth et al. 2009, p. 21). Furthermore, the previous

[7]Following Nölke, transnational networks are relevant in situations where the interests among the actors are homogeneous and where there is little institutionalization in the national arena (Nölke 2000, p. 335). Other prominent representatives that can be subsumed under the umbrella term 'transnational network research' are Haas' (1990, p. 58) 'epistemic community' approach and Tallberg's et al. (2013) research on the relationship between IOs and transnational actors.

[8]For a historic overview of reform processes in public administration, see Mehde (2007).

and predominant focus on the questions of IO creation and what role organizations play in general has been revised, and the more recent research on IOs has instead focused on the performance and effects of IOs (Abbott et al. 2015; Ege 2016; Hall et al. 2015; Pease 2016; Rosenow-Williams and Gemenne 2016). Building on this research, the following sections amplify the theoretical inputs that will guide the research design of this thesis. The first section outlines the aspects of security theory that help to capture the security references used by the UNHCR in the context of EDPs. The second section elaborates on how organizational theory contributes to this research project.

#### 1.3.2.1 Input from Security Theory

First, the public, political, media, and scholarly representations of environmental changes in the 2000s have distinctively interlinked the topics of EDPs and security (Kita and Raleigh 2018; McLeman and Gemenne 2018a). According to Kita and Raleigh, "there is a growing public narrative that presents environmental change as a source of political instability and conflict within developing nations, and which presents international migrants from developing nations as a potential source of international security risks" (Kita and Raleigh 2018, p. 357). International actors like the UNHCR do not exist in a vacuum, which means that the organization did watch this perception of EDPs in the international realm. Under the leadership of High Commissioner Guterres (2005–2015), the UNHCR started to pick up the topic of environmental changes and the organization distinctly used security references as regards EDPs (see Sects. 1.1.2 and 5.2; Hantscher 2016). These developments until 2015 explain why this thesis relies on input from security theory. Put differently, it needs a theoretical approach that helps to grasp the UNHCR's interest in EDPs and the accompanying changes in the UNHCR's debates on and depiction of the topic in terms of security.

Security theory is therefore essential, as it allows a first glimpse into a possible organizational change inside the organization. This thesis assumes that the UNHCR's conception of EDPs becomes transparent in the High Commissioner's speeches and the UNHCR's publications, websites, and documents, because these sources illustrate what the organization thinks about the EDP matter and what kind of image it wants to convey to the outside world. The attention of the organizational analysis then shifts to the UNHCR's DNA and the organization's general ideals. That is why this thesis uses the term 'ideational depiction' when referring to the hints in the above-mentioned sources that mirror the UNHCR's stance on the topic of EDPs. From 2005 onwards, it shows that the UNHCR's conception of EDPs has revolved around and incorporated different security references that present themselves predominantly in the organization's official statements and publications (see Sect. 5.2). Nevertheless, what is understood under the term 'security' varies: security sometimes implies the security of a specific state (national security), the security of the international community (international security), or the security of the individual who suffers from the impacts of environmental changes (human security).

In order to be able to grasp the different security references in the UNHCR's approach toward EDPs, this thesis relies on and applies the 'tools' of security theory. Security theory has its roots in, and is a sub-theory of, International Relations theory (IR). The research design of this thesis breaks down those theoretical 'lenses' of security theory that are the most helpful in answering the research question at hand. The following sections will therefore introduce some of the background to security theory along with the theoretical strands of the Copenhagen School, the Paris School, and the Human Security concept (see Sect. 2.1).

**The Copenhagen School and the Paris School**

The first reason for choosing the Copenhagen School and the Paris School is that the current security research mainly springs from these two schools of thought (Bourbeau 2014, p. 187); second, the special emphasis of these theoretical 'lenses' is essential to be able to approach the empirical puzzle. The Copenhagen School and the Paris School tackle the question of whether it is advisable to address issues in all kinds of sectors—e.g., the military, environmental, economic, societal, and political sectors—in security terms (Buzan et al. 1998). Therefore, security references in the environmental realm lie close to the schools' vantage points. Both schools thus allow for critical reflections on security references, such as the UNHCR's depiction of EDPs.

Buzan, Wæver, and de Wilde of the Copenhagen School introduce the term 'securitization' which aims, firstly, to describe the process by which a specific issue is being put into a security context; secondly, to identify by whom; and, thirdly, to outline with what effects (Buzan et al. 1998, p. 26). By this means, the Copenhagen School helps to identify those who are threatened, and it uncovers the actors who implement the securitizing move (Jackson 2006, p. 301). Moreover, the theory allows examination of how a securitization process unfolds by shifting the focus to security references. For the context of this thesis, a successful securitization of EDPs would thus imply that an actor depicts EDPs in national or international security terms—e.g., as threats to national sovereignty or international peace and security. In addition, the ideational depictions would need to include traditional security responses, such as the involvement of the military. Lastly, the protection of EDPs would lead to extraordinary measures that the public normally perceives as unacceptable (see Sect. 2.1).

The sociological approach of the Paris School essentially extends the argument of the Copenhagen School to non-discursive practices. Additionally, the Paris School considers the institutions of the securitizing actor, and thus aims to unveil the power positions and institutional struggles behind a securitizing act. This theoretical 'lens' thereby allows better insight into the institutions behind the person who securitizes an issue (Bigo 2002, p. 74). In the context of environmental changes, the Paris School assumes a 'climatization' of the security field—i.e., professionals in security paint a picture of environmental changes as threats to their daily business (Oels 2012, p. 185). Examples of this climatization are the military, the police, or the defense ministries that have practically included environmental changes in their early warning systems and planning studies (Oels 2012, p. 185).

**The Human Security Concept**

Building on how the Copenhagen School and the Paris School look at the matter of environmental migration and displacement, it is seen that these theoretical perspectives suffice to grasp the usage of security references. Nevertheless, the UNHCR's humanitarian outlook influences the organization's depictions, which means that the people of concern are the primary object of reference for the UNHCR. Therefore, the organization's publications and official speeches depict that the UNHCR is very clear in its terminology and directly refers to refugees, IDPs, or EDPs. Despite the usefulness of the theoretical approaches of the Copenhagen School and the Paris School in terms of reflecting on the use of security references in general, the two schools' theoretical outlook does not put the human center stage. In other words, the security theories mentioned above do not provide all the theoretical background required for the empirical puzzle of this thesis.

As the Copenhagen School and the Paris School lack a human-centeredness, an additional theoretical concept needs to complement their contribution to this thesis. This security theory needs to fulfill the criteria of comprehending and grasping the UNHCR's depictions that refer (a) to individuals and (b) to their inherent need for security. To explain the UNHCR's depictions of EDPs, it is therefore necessary to introduce the concept of Human Security. Applied to the topic of environmental changes, human security would imply a shift in the policy agenda toward sustainable development (Oels 2012, p. 185). In addition, a human security concept of EDPs would imply a shift in awareness toward, first, those individuals who have suffered because of environmental changes and, second, the way people have tried to cope with (sudden or long-term) changes in the environment.

'Security' is a controversial term. Security has a variety of meanings, as it is "a societal value, a political goal, and an instrument aimed at protection, risk reduction, certainty, predictability, reliability, trust and confidence, in contrast to danger, risk, threat, disorder and fear" (Scheffran et al. 2012, p. 3). Alexander Wendt, on the one hand, asserted that security incorporates an intersubjective meaning and can be made by actors (Wendt 1992); Buzan, Wæver, and de Wilde, on the other hand, referred to security as a situation in which problems are upgraded to the highest importance, thus requiring extraordinary measures (Buzan et al. 1998). The advancement of the term toward human security implies a shift in the referent object from the security of states (including their respective national security conceptions) to a people-centered approach in which humankind has become the referent object (Scheffran et al. 2012, p. 3).[9]

The application of the Copenhagen School, the Paris School, and the Human Security concept in this thesis allows grasping of the UNHCR's depictions on the matter of EDPs. Nevertheless, this perspective on the empirical puzzle alone does not suffice, because the theoretical 'lenses' of security theory only shed light on the depictions that the UNHCR uses in its publications and speeches on EDPs. In other words, the 'tools' of security theory only look at the organization's EDP approach by

[9]For more on the Human Security concept, see Sect. 2.1.3.

means of the UNHCR's own wording and by interpreting the UNHCR's ideals and norms, which might influence the approach. Therefore, this thesis uses security theory to explain the empirical puzzle and to capture the changing security references of the UNHCR from 2005 until 2015. Nevertheless, the empirical puzzle also requires a theoretical perspective that can explain the push and pull factors that draw the UNHCR toward the topic of EDPs. The theoretical 'lenses' originating in organizational theory fill this gap on the 'why' question and the next section will therefore introduce the input from organizational theory.

#### 1.3.2.2 Input from Organizational Theory

Second, organizational theory is essential to answer the question of why the organization acts the way it does. The assumptions deduced from organizational theory thus shed light on the 'black box' of the organization. Organizational theory originates in sociology, and the application of organizational theory in IR is a rather recent development of the last decades (see Sect. 3.2).

The theoretical approach of this thesis responds to the call of scholars to bridge the rationalism and constructivism divide by pragmatically combining theoretical approaches that shed light on different elements of the UNHCR's behavior (building upon and similar to Gutner 2005, p. 14; Hanrieder 2014a, p. 324; Nielson et al. 2006, p. 115). As argued by Börzel and Risse in their proposed synthesis of rationalism and constructivism, meta-theoretical borders often need to be crossed for empirical analyses; when analyzing the real world, scholars should not be forced to decide on one logic of action (Börzel and Risse 2001, p. 12). Rather, Börzel and Risse advocate that scholars explore which logic of action, or which theoretical background, dominates in the respective empirical case. Drawing on Nielson et al. (2006, p. 114), this research design aims to build upon the differing hypotheses from organizational theory concerning organizational behavior and change. Hence, this thesis criticizes the perception that one has to decide whether internal or external factors capture the organizational processes of the UNHCR best. Instead, the thesis responds to Siebenhüner's (2008, p. 93) and Weaver and Leiteritz's (2005, p. 370) appeal that it requires a focus on both extra-organizational factors and intra-organizational features to fully understand organizational behavior and change.

Therefore, the empirical puzzle of this thesis needs a theoretical approach that incorporates the influences of the organizational environment, the member states, the UNHCR's bureaucracy, and the organization's goals and strategies regarding EDPs. This list of possible factors implies that the UNHCR's approach toward EDPs is a puzzle for the individual strands of organizational theory, which explains why this thesis proposes an all-encompassing theoretical approach. As a result, it builds upon Barnett and Finnemore's perspective as outlined in their 2004 work *Rules for the World: International Organizations in Global Politics*. Barnett and Finnemore's theoretical understanding is to "make IO interests a subject of inquiry rather than assuming them and mak[ing] bureaucratic culture central to the investigation" (Barnett and Finnemore 2004, p. 159).

Nevertheless, the thesis goes even further than what Barnett and Finnemore proclaim, because it combines the fields of research on intra-organizational and extra-organizational factors. In the former, the respective IO itself is the analytical unit and the researcher focuses on how political decisions come about inside the organization (e.g., inside the UNHCR), taking into consideration the organizational culture or bureaucratic styles (Enkler and Grohs 2017, pp. 255–258). The latter, the external organizational approach, has one specific policy field as the analytical unit (e.g., the refugee regime) and the researcher focuses on how effective an IO acts in this policy field (Eckhard and Mattmüller 2017, p. 230) or how IOs collaborate in this policy field (Rittberger and Zangl 2003, p. 89). In other words, this thesis examines both the UNHCR as the analytical unit and the way that the UNHCR conducts itself in the refugee regime. In doing so, it takes both intra-organizational and external factors into consideration.

To sum up, the theoretical approach of this thesis ensures that the assumptions derived from organizational theory support the ideational depictions from security theory. The organizational factors cover the UNHCR's organizational environment and the role of member states, but also intra-organizational factors, such as the bureaucracy and organizational strategies. The assumption is that the UNHCR's approach toward EDPs might arise from one of these factors, from a combination of these factors, or from none of these factors (see Sect. 4.2). The combination thus illuminates the pathway of the UNHCR's approach (see Fig. 1.1). By adding the security references to the assumptions from organizational theory, the thesis will explain the underlying *hidden* organizational agenda, while taking the UNHCR's *visible* and externally-oriented depictions of EDPs into account. Details on the mutual benefits of security theory and organizational theory are given below.

#### 1.3.2.3 Setting Up Ideational Depictions and Organizational Factors

The ideational depictions are broadly informed by security theory in a case-related (UNHCR) manner. Similarly, the organizational factors derive from organizational theory in a case-related (UNHCR) manner. The literature review shows that empirical case studies have not so far combined these two theoretical backgrounds (see Sect. 3.3). Nevertheless, this theoretical combination is central to the underlying research question of this thesis.

Figure 1.1 depicts the combination of the two perspectives and thereby presents the conceptual design of the analysis of this thesis. The combination of the ideational (left) and organizational factors (right) allows explanations of the UNHCR's EDP approach and taking the organization's concepts, structures, hierarchies, and personality traits into consideration. The conceptual design does not aim to play the two backgrounds off each other; rather, the theoretical approach of this thesis builds upon elements of both perspectives for different purposes. By choosing the case-related and most relevant aspects of both, the concept aims to answer the overall research question. Here, the theoretical background of security theory helps to grasp the empirical puzzle. In 2005, the UNHCR began to use security references in the

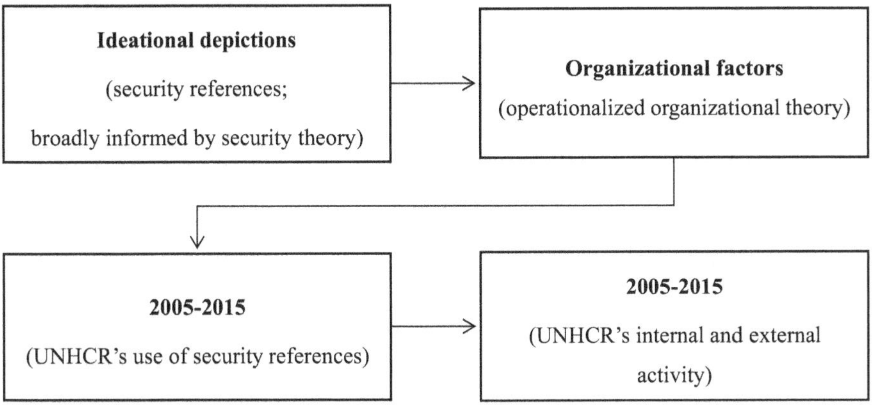


**Fig. 1.1** Combination of theoretical perspectives

context of environmental changes (see Sects. 1.1.2 and 5.2), and this organizational behavior is especially noticeable until 2015. Therefore, the analysis of the empirical puzzle at hand relies on security theory to illuminate the UNHCR's depictions. Adding to this, organizational theory will shed light on organizational aspects of the UNHCR. In this context, the analysis will closely review the UNHCR's internal workings as well as the external organizational activity from 2005 until 2015.

By setting up the two theoretical outlooks, this thesis therefore does not aim to 'build' a new strand of security or organizational theory; instead, the thesis uses the different theoretical perspectives as a 'toolbox' to address the empirical puzzle. That said, only the combined approach of ideational depictions and organizational factors sufficiently answers the research question of why the UNHCR approaches the matter of EDPs the way it does. Therefore, the next section provides an insight into the methodology and the conceptual design of this thesis (also see Chap. 4).

### *1.3.3 The Methodology*

In a first step, the thesis converts the case-related information from security and organizational theory into a list of theoretical assumptions about the UNHCR's EDP approach (see Sect. 4.2). The chapter of methods condenses aspects from the current state of security and organizational research, while the analysis applies the main assumptions from both research strands to the UNHCR's decision to approach the topic of EDPs (see Sects. 4.1 and 4.2).

*Step 1: Developing **theoretical assumptions** from security and organizational theory*

The second step is the operationalization of the theoretical assumptions. This includes establishing a list of questions that subsumes the essential elements of the first step. These questions aim at building bridges between the theoretical assumptions and the empirical case study. Building upon similar approaches in existing scientific research (among others, Barnett and Coleman 2005, p. 593; Barnett and Finnemore 1999, p. 726; Biermann et al. 2009b; March 1989, p. 170; Jetschke and Theiner 2016, p. 7), this comprehensive approach intends to contribute to a better understanding of the behavior of the UNHCR.

*Step 2:* ***Operationalizing*** *the theoretical assumptions into a list of questions*

The third step transfers the deduced question areas into an all-encompassing theoretical concept for the thesis—that is to say, the approach combines the list of questions deduced from organizational research with the assumptions derived from security theory. The overall theoretical concept will then answer the central research question of why the UNHCR approaches the issue of EDPs the way it does, and will do so by shedding light on both the depictions used by the UNHCR on the topic (security theory) and the organizational factors that might have triggered its approach (organizational theory).

*Step 3: Transferring the list of questions into an* ***overall theoretical concept***

Table 1.1 shows that the analysis tackles the UNHCR's approach during the time period of 2005 to 2015 (see Sect. 4.2). For this timeframe, the analysis will carve out the UNHCR's ideational depictions—i.e., the security references—and the thesis will examine if the UNHCR's conception of EDPs changes over time. In addition, the analysis will reassess whether the organizational factors apply to the organization's interest in EDPs from 2005 until 2015. In terms of its contribution to organizational research, the conceptual design advocates the importance of in-depth research on IOs and, here especially, the kind of research that acknowledges IOs as both agents and principals (similar to Schroeder 2014, p. 357). The theoretical concept suggests that in the case of an empirical puzzle such as that of the UNHCR, the research design must consider organizational factors from both macro- (e.g., organizational environment and member state involvement) and micro-perspectives (e.g., bureaucracy and organizational strategies). The analysis of this thesis will therefore assess these diverse organizational factors one after the other and ultimately evaluate the influence of the respective factor. By combining the macro- and micro-perspectives, the conceptual design responds to the research gap identified by, among others, (Barnett and Coleman 2005), and thus has implications for organizational research and research on IOs.[10] Concerning the various organizational perspectives of Table 1.1, there certainly are overlaps and agreements between the theoretical perspectives. For example, although this thesis describes the theoretical strands of intra-organizational factors and the character of the bureaucracy separately, they examine similar internal matters of IOs. The theoretical 'lenses' thus interact and the theoretical approach of this thesis only separates them to gain analytical clarity.

[10]The thesis uses the term 'research on IOs' for research on international organizations and the term 'organizational research' for research on organizational theories.

**Table 1.1** Conceptual design for the empirical analysis of this thesis

| | Ideational depiction (security references; broadly informed by security theory) | Organizational factors (operationalized organizational theory) | | | |
|---|---|---|---|---|---|
| | | Organizational environment | Member states | Character of the bureaucracy | Inner-organizational strategies |
| 2005–2015 | 'Traditional' security /human security | Relevant/not relevant | Relevant/ not relevant | Relevant/ not relevant | Relevant/not relevant |

#### 1.3.3.1 Determining the Text Corpus

The thesis at hand applies Philipp Mayring's method of qualitative content analysis to the overall theoretical concept. Thereby, the theory-led categories constitute the road map for the analysis of empirical data. In Mayring's words: "Theoretical orientation means, then, the tapping of this experience [of others in the same field] in order to achieve an advance in knowledge" (Mayring 2014, p. 59). Mayring also specifies which types of material fit his model of qualitative content analysis. He mentions interview transcripts, documentations in institutions, webpages, protocols, and databases of institutional texts (Mayring 2014, p. 43). This research project orients itself with Mayring's examples and therefore uses the following corpus of documents for the analysis.

As primary sources, this thesis uses UNHCR speeches and publications, information from the organization's website, and articles from other IOs on the topic of the UNHCR. By examining the UNHCR's depiction of EDPs, possible conclusions can be inferred about the UNHCR's underlying intentions. Adding to the literature research, the analysis is based upon qualitative expert interviews. The aim of the interviews is, first, to double check the information received from the primary and secondary sources and to add value to them. Second, the qualitative expert interviews allow a better view into the organization and into the inner workings of the UNHCR, as compared to the official publications that are available on the organization's website. The information gained from the expert interviews thereby contextualizes the other data and adds value to the information found in the UNHCR's publications. The secondary sources of the analysis are academic publications on the topic of the UNHCR and on the organization's EDP approach. All of the material is written text and thus lends itself for a qualitative content analysis (see Sect. 4.3.2 and cf. Mayring 2014, p. 57).

#### 1.3.3.2 Qualitative Expert Interviews

The thesis approaches the empirical puzzle by a combined analysis of primary and secondary sources. Because the primary sources—i.e., organizational publications, etc.—only shed light on one side of the organization (the outer façade), the analysis has to incorporate a glimpse inside the organization to uncover the inner workings of the UNHCR. Therefore, people who work at the UNHCR and people who work together with the UNHCR staff are paramount to the analysis. The impressions from the expert interviews will help to rate the findings from the primary and secondary sources and therefore add value to the analysis. The expert views thicken the interpretation and put the primary and secondary data into context. That said, the results from the interviews do not constitute the only database; rather, the results represent additional interpretation devices.

Following previous research on IOs (cf. Biermann et al. 2009b), this thesis approaches the validity of interview data and the bias problem in a twofold manner. First, the interviewer intends to create a confidential and easy atmosphere; second, the experts are interviewed one at a time, without additional staff in the room. Furthermore, this thesis approaches the criteria for the validity of the research design with the help of a triangulation of methods. In social science, triangulation sometimes refers to the mix of qualitative and quantitative methods, but there exists a second meaning of the term: a triangulation of different qualitative methods to interpret the existing data and to minimize the disadvantages of one specific method (cf. Blatter et al. 2007, p. 37) Gläser and Laudel 2004, p. 102). Therefore, this thesis uses different methods to guarantee validated results (see Chap. 4). Another central research criterion is the intersubjective verifiability. This implies that the researcher precisely documents the whole research process as well as the data that he or she collects (Westle 2009, p. 259). The thesis at hand fulfills this criterion by describing the whole research process (see Chap. 4) and adding the interview transcript to the thesis (see attachments).

**Choice of Experts**

Overall, this research project factors a dozen expert interviews into the analysis, with staff members from the UNHCR, the IOM, the NRC, the Nansen Initiative, and the IDMC, as well as academics and several state representatives[11] who have been involved in the topic. Previous organizational research that also included different types of experts (e.g., Biermann et al. 2009b; Rosenow-Williams 2016) inspired the broad range of experts for the interviews. Additionally, the career paths of the experts were taken into consideration. According to Kaiser (2014), the person who holds the highest position in an organization is not necessarily the one with the best overview of the internal workings. Therefore, all the experts have different job functions and different years of experience, from junior professionals to senior staff members. The decision to contact an expert was based solely on his or her knowledge about the UNHCR's approach toward EDPs. This explains the number of expert interviews for this thesis: usually, there are only one or two experts on a certain topic inside an organization (Kaiser 2014, p. 133). Therefore, during the field

[11]The choice of state representatives in Geneva is intended to include representatives from both developed and developing countries. In other words, the intention was also to talk to representatives from those countries that experience migration and displacement due to environmental factors. The experts from developed countries were chosen because of their respective country's involvement in the matter of environmental migration and displacement.

research in Geneva, it was not difficult to find interview partners; rather, the challenge was to get the 'right' people—those who are the experts on the topic and who were willing and able to do an interview.

**Data Analysis**

The data analysis is first based upon a topic-related literature review of primary sources and secondary sources on the UNHCR and then on a software-based qualitative content analysis of a dozen semi-structured qualitative expert interviews. The qualitative content analysis is a theory-led method of analysis that aims to structure and evaluate the content of texts (cf. Blatter et al. 2007, p. 34). The first specific trait of content analysis is that the method extracts information from a text and then works with this condensed and shortened new text. This procedure opposes other methodological approaches that mostly work with the original text material (Gläser and Laudel 2004, p. 44). The second specific trait of content analysis is that the scientist who applies this method builds the category system prior to the textual analysis (Gläser and Laudel 2004, p. 44). Overall, the centerpieces of content analysis are the categories and the coding processes (Kuckartz 2009, p. 336). For the software-based qualitative content analysis of the expert interviews, a coding system is used.

The overall aim of the coding and categorizing of the interview transcripts is that "parts of the data (e.g., a statement) are taken out of their context and grouped with other bits of data (similar statements) and put in relations—categories, subcategories, etc." (Flick 2014, pp. 44–45). Based on previous qualitative research with expert interviews (cf. Flick 2014; Rosenow-Williams 2016), the analysis uses the software program MAXQDA. The coding of the qualitative expert interviews orients itself with the coding method developed by Kaiser, whose approach constitutes a well-established and straightforward method that aims at the content without further ado. This thesis therefore follows the definition of coding by Flick, who argues: "Coding is the work with materials for generating concepts and for allocating excerpts of the material to categories. This is then followed by interpretation: how can we understand and explain what the coding has revealed about the views in the field, their links among each other, their ties to context conditions, and the like" (Flick 2014, p. 373).

## 1.4 The Empirical and Theoretical Contributions of the Thesis

By bringing together the information on the case, the outline of the empirical puzzle, and the approach of the research project, this research project depicts relevance on both an empirical and a theoretical level. The following two sections substantially add to this introductory chapter and lay the groundwork for the subsequent chapters as regards the 'so what' question.

### 1.4.1 The Societal Relevance of the Project

The societal relevance of a project means the valuable contribution that it makes not only to the scientific community but also to the public. In other words, "a research project should pose a question that is 'important' in the real world" (King et al. 1994, p. 15). Societal relevance implies that a research helps to understand a political phenomenon that concerns the well-being of a certain group of people; ideally, the project informs the public about a certain political process and thus adds to the general education (Lehnert et al. 2007, p. 41). It is important whether the public acknowledges the relevance of research, and whether there exists a personal attachment to the issue at hand—that is, whether political developments concern the people (Lehnert et al. 2007, p. 45). The empirical puzzle for a research project must thus seem plausible and relevant to the lives of the concerned public (see Sects. 6.2 and 6.3). Through academic research, the political and social processes become central, and the respective scientists consider how these processes influence everyday life. By applying these criteria of societal relevance to this thesis, it is possible to make out the group of people that suffers from environmental changes, or will suffer in the future.[12] Thereby, the research question touches upon the debate surrounding the status of EDPs and the question of whether EDPs become accepted refugees or stateless persons with no allowances, or whether states will succeed in creating a new category of protection for EDPs. The way the UNHCR responds to EDPs therefore directly affects those people who must migrate due to environmental changes. That said, the UNHCR's actions ultimately influence the status of many people, today and in the future.

International migration and refugee flows directly involve the topic of EDPs; it has thus become paramount to pay attention to those global actors who cover the fields of refugees and migration. IOs like the IDMC, the IOM, the NRC, and the UNHCR largely lead and control the field of migration and refugees. However, the flow of refugees and migrants also directly affects the national interests of states, and it falls under the jurisdiction of states to guarantee refugee and asylum rights. This makes the question of whether the UNHCR decides to help and assist EDPs or not highly relevant. There exists a discrepancy in this field of international politics that is organized and led by IOs on the one hand, while largely affecting state interests on the other hand. Research projects with a similar research interest to this thesis have already stressed the academic relevance of such endeavors for understanding global governance; that said, the benefits lie in shedding light on the cooperation among IOs, between IOs and societal actors, and between IOs and states (cf. International Public Administration DFG Research Unit 2018).

[12]The reason for there being so few commonly accepted predictions on the probable number of EDPs stems from the lack of definitions of the terms 'environmental/climate change migration' and 'environmental/climate change refugee' (Warner 2010, p. 403) up until the launch of the Nansen Initiative in 2012.

Adding to this, the societal relevance of the empirical puzzle of this thesis shows in the current debates on the topic in the media, among politicians, and in civil society. This is not to say that a research project should be chosen due to its current appeal, e.g., by the buzzwords 'climate change' (Lehnert et al. 2007, p. 49). Yet, a certain present-day importance implies that the public perceives the topic of EDPs as relevant and worthwhile to talk about. The chosen empirical puzzle exemplifies to what extent scientific research can have practical implications, i.e., to what extent practitioners pick up the topic. For the case study of the UNHCR, this means that the relatively small group of people who work on the topic—the scientific community and the civil servants from the IOs—are in close contact and repeatedly discuss the matter in meetings, at conferences, and during workshops. Therefore, IOs acknowledge what scientists recommend and sometimes even consider the academic input in their decision-making and plans. This mutual stimulation has shown during the last 5 years at the various conferences and workshops on the topic of EDPs, at which the academic community and the staff of IOs has participated and discussed each other's work.

### *1.4.2 The Theoretical Relevance of the Project*

As regards the thesis at hand, the research design illustrates theoretical relevance. The theoretical relevance refers to the value that a given research project adds to either the scientific discussion or the development of theory. King, Keohane, and Verba describe this type of relevance as follows: "A research project should make a specific contribution to an identifiable scholarly literature by increasing our collective ability to construct verified scientific explanations of some aspect of the world" (King et al. 1994, p. 15).[13]

In IR, organizational theories have only been used in order to add certain elements to existing IR theories (see Sect. 3.1). Therefore, organizational approaches are simply used to get a better insight into organizational change (Dingwerth et al. 2009, pp. 18–19). IR scholars largely treat IOs as being of little importance when it comes to explaining the world; i.e., they have "rarely investigated the extent to which IOs have acquired autonomous influence over the formulation of international policies" (Verbeek 1998, p. 11). The underlying reasoning is that the two theoretical strands ask different kinds of research questions: whereas organizational research uses organizations as the dependent variable that needs to be explained, political science makes organizations the independent variable to explain power

[13]More specifically, the aim should be to relate the planned project to the already existing literature on the topic and to put a research project in the context of previous scientific literature on the topic (King et al. 1994, p. 16). A scholar can give theoretical relevance if he or she: (1) empirically tests previously untested theoretical hypotheses; (2) uncovers inconsistencies in a theory; (3) identifies empirical cases which cannot be explained by the existing theories; or (4) formulates alternative explanations for an empirical observation (Lehnert et al. 2007, p. 43).

constellations (Dingwerth et al. 2009, pp. 18–19). An explanation for the recent rise of interest in organizational theory is that IR theories struggle to grasp the internal motivations, problems, and agendas of IOs (Hanrieder 2014b, p. 221). Questions of the legitimacy and accountability of IOs have taken center stage in much of the current organizational research (Kleine 2013, p. 325), and the research on IOs as arenas and the interest in IOs as bureaucracies becomes clear from this background (Balint and Knill 2007, p. 117).

First, the 'black spots' of the grand theories of IR (see Sect. 3.1) motivate the search for a theoretical approach that is able to look into both the external and the internal factors which might have triggered the UNHCR's interest in EDPs. The thesis at hand thus exhibits theoretical relevance because the empirical puzzle originates in IR, but the research design nevertheless leaves behind the worn-out theoretical lenses of IR. Instead, this research moves toward organizational research. Thereby, this thesis aims to surpass the broad separation of, first, the IR approach to IOs and, second, the public administration approach to IOs. As Ege and Bauer put it, "IR usually is better in explaining why IOs are created, whereas Public Administration is better suited to analyzing the policy-making role of IOs and their bureaucracies in day-to-day politics" (Ege and Bauer 2013, p. 135). That is why the work at hand builds a bridge between the two approaches (IR and public administration). The research design of this thesis settles in this theoretical gap and thereby tackles the problem of applying IR theory in an IO context.

Second, the thesis does not shy away from combing organizational theories with security theory from IR. The empirical puzzle—i.e., the UNHCR's interest in a matter outside its mandate—has so far not been satisfactorily grasped in theoretical terms. That is why the chosen theoretical approach takes a broader perspective when it comes to choosing a theoretical basis. To approach the empirical puzzle, it needs input from security theory (see Sect. 4.1). Various security perspectives exist on the topic of environmental migration, starting with the Copenhagen School and moving on to the Human Security concept. In the primary and secondary data of the UNHCR's EDP approach, a change in rhetoric and action is apparent. Chapter 5 will give details first on how the Copenhagen School best explains the security references by the UNHCR until around 2010, and then to what extent the UNHCR's depiction of the topic showed a Human Security conception of EDPs until 2015. Therefore, the Copenhagen School, the Paris School, and the Human Security concept are helpful when considering the empirical data, and all are relevant for explaining the empirical puzzle.

Summing up the societal and theoretical relevance, the benefit of this thesis is of a twofold nature. Firstly, the thesis presents explanations for an empirical puzzle—that is, the UNHCR's interest in EDPs—and thus adds to a better understanding of global governance in the twenty-first century. Secondly, the work at hand has theoretical value, because it establishes a relationship between two previously unrelated factors (namely ideational and organizational factors), and thereby contributes to current security and organizational theory.

## 1.5 The Structure of the Chapters

Building upon this introductory chapter, Chap. 2 establishes the case and outlines the security connotations with regard to EDPs. Chapter 2 introduces some of the leading security theories that serve as a background for many of the scholarly approaches to the topic. By this means, Chap. 2 shows the development from an environmental conflict debate toward the current debate on climate change and natural disasters in terms of human mobility. In this way, the chapter presents a background on EDPs and places the topic within the context of migration studies, displacement studies, refugee studies, and security studies. Chapter 2 therefore lays the groundwork for the analysis.

Chapter 3 represents the theory chapter—i.e., the chapter discusses the applicable theoretical background from organizational research. In this way, Chap. 3 provides a basis for the theoretical approach of this thesis. After a brief chronological overview of the state of the art of organizational research, the focus shifts to a more theory-led overview of organizational research. The chapter thereby compiles the central theoretical assumptions of the major branches of organizational research. The intention is to extract those aspects of the respective sub-theories of organizational research that help to elucidate the empirical puzzle. Accordingly, by looking at the different theoretical 'lenses' of organizational research, the theory chapter asks the questions 'how does the respective sub-theory of organizational research explain the UNHCR's approach toward EDPs?' and 'what can the respective theory contribute to the organizational analysis?'

Building upon this, Chap. 4 elaborates on the research design of the thesis and builds the overall conceptual design that will be applied to the UNHCR. The chapter introduces the chosen qualitative methods and explains the purpose of these methods in the research process of this thesis. Thereby, Chap. 4 gives details on the research process and illustrates how the inputs from Chap. 3 help to develop the guided interview. For matters of comparability of data collection, the chapter reveals the set of questions—i.e., the questionnaire for the interviews—and explains the specifics of a half-standardized interview, a method that is both deductive and inductive (Biermann et al. 2009b, p. 62). After a recapitulation of the interview situation in Geneva, the chapter outlines the transcript rules of the expert interviews, presents the steps of the qualitative content analysis, and sheds light on the coding of the interview material. By these means, the research project at hand fulfills the criterion of intersubjective verifiability mentioned in the scientific literature (Westle 2009, p. 259).

Chapter 5 is the chapter of analysis. The chapter has a chronological structure: it starts with the beginning of Guterres' reign in 2005 and ends with the year 2015. In addition, the analysis develops according to the four strands of organizational research that were compiled in Chap. 3. That said, Chap. 5 has four sections, each of which focuses on the empirical puzzle from one theoretical perspective. The first section looks at how the organizational environment might have influenced the UNHCR's EDP approach. It thereby discusses the influence of international politics as well as the influence of other IOs. The second section depicts the role of member states in the UNHCR's interest in EDPs. It thus builds on organizational theories that have a principal-agent perspective on IOs. The third section takes an intra-organizational perspective and focuses on the character of the bureaucracy—i.e., the UNHCR staff and the High Commissioner António Guterres. The fourth and final section builds upon organizational research which expects that organizational goals or organizational window-dressing might play a role in the UNHCR's approach. Therefore, this section elaborates on the influence of diverse intra-organizational strategies which make assumptions about the empirical puzzle. As a result, all four sections implement the conceptual design of Chap. 4. In order to guarantee anonymity, the analysis does not include the names of the interviewees, which is a common procedure in situations of interviews with a manageable group of IO staff (cf. Biermann et al. 2009b, p. 63).

Lastly, Chap. 6 concludes by assessing the findings of the analysis. The chapter first recapitulates the findings in accordance with the four sections of the analysis. Chapter 6 then discusses the findings and classifies them within the existing research on IOs and the broader context of global governance. Chapter 6 also includes a critical reflection of this thesis' research design—i.e., the chosen theoretical concepts and the chosen methodology. Building upon this conclusion, the chapter also gives an outlook which comments on both the UNHCR and the topic of EDPs.

To conclude this chapter laid the groundwork for the following chapters by introducing the topic of disaster displacement and the way the UNHCR has looked at the issue so far. By this means, this introductory chapter has outlined the empirical puzzle as well as the existing research gap on the UNHCR's EDP approach. Additionally, this chapter depicted the approach of this project, including the case selection, the theoretical approach, and the methodology. Lastly, the introduction illustrated how the thesis contributes to empirical and theoretical research and it gave an overview of the subsequent chapters. Chapter 2, constitutes the next step as it establishes the case and sheds light on how migration studies, disaster displacement studies, refugee studies, and security studies have approached the topic of EDPs.

## References

Abbott KW, Genschel P, Snidal D, Zangl B (eds) (2015) International organizations as orchestrators. Cambridge University Press, Cambridge

Baldwin A, Bettini G (2017) Introduction: life adrift. In: Baldwin A, Bettini G (eds) Life adrift: climate change, migration, critique. Rowman & Littlefield, Washington, DC, pp 1–22

Balint T, Knill C (2007) The limits of legitimacy pressure as a source of organizational change: the reform of human resource management in the OECD. In: Bauer MW, Knill C (eds) Management reforms in international organizations, vol 6. Nomos, Baden-Baden, pp 117–132

Barnett M, Coleman L (2005) Designing police: interpol and the study of change in international organizations. Int Stud Q 49:593–619

Barnett MN, Finnemore M (1999) The politics, power and pathologies of international organizations. Int Org 53:699–732

Barnett M, Finnemore M (2004) Rules for the world: international organizations in global politics. Cornell University Press, New York

Barnett J, Webber M (2010) Migration as adaptation: opportunities and limits. In: McAdam J (ed) Climate change and displacement. Hart, Oxford, pp 37–56

Bauer S, Biermann F, Dingwerth K, Siebenhüner B (2009) Understanding international bureaucracies: taking stock. In: Biermann F, Siebenhüner B (eds) Managers of global change: the influence of international environmental bureaucracies. MIT Press, Cambridge, MA, pp 15–36

Betts A (2011) Introduction: global migration governance. In: Betts A (ed) Global migration governance. Oxford University Press, New York, pp 1–33

Biermann F, Siebenhüner B (2009) The role and relevance of international bureaucracies: setting the stage. In: Biermann F, Siebenhüner B (eds) Managers of global change: the influence of international environmental bureaucracies. MIT Press, Cambridge, MA, pp 1–14

Biermann F, Siebenhüner B, Schreyögg A (eds) (2009a) International organizations in global environmental governance. Routledge, New York

Biermann F, Siebenhüner B, Bauer S, Busch P-O, Campe S, Dingwerth K, Grothmann T, Marschinski R, Tarradell M (2009b) Studying the influence of international bureaucracies: a conceptual framework. In: Biermann F, Siebenhüner B (eds) Managers of global change: the influence of international environmental bureaucracies. MIT Press, Cambridge, MA, pp 37–74

Bigo D (2002) Security and immigration: toward a critique of the governmentality of unease. Alternatives 27:63–92

Black R, Kniveton D, Schmidt-Verkerk K (2011a) Migration and climate change: towards an integrated assessment of sensitivity. Environ Plann A 43:431–450. https://doi.org/10.1068/a43154

Black R, Adger WN, Arnell NW, Dercon S, Geddes A, Thomas D (2011b) The effect of environmental change on human migration. Glob Environ Change 21:3–11. https://doi.org/10.1016/j.gloenvcha.2011.10.001

Blatter JK, Janning F, Wagemann C (2007) Qualitative Politikanalyse: Eine Einführung in Forschungsansätze und Methoden. VS Verlag für Sozialwissenschaften, Wiesbaden

Boas I, Rothe D (2016) From conflict to resilience? Explaining recent changes in climate security discourse and practice. Environ Polit 25:613–632

Börzel T, Risse T (eds) (2001) Die Wirkung internationaler Institutionen: Von der Normanerkennung zur Normeinhaltung: Working Paper. Reprints aus der Max-Planck-Projektgruppe Recht der Gemeinschaftsgüter 15

Bourbeau P (2014) Moving forward together: logics of the securitisation process. Millennium J Int Stud 43:187–206. https://doi.org/10.1177/0305829814541504

Buzan B, Wæver O, Jd W (1998) Security: a new framework for analysis. Rienner, Boulder, CO

Castles S (2010) Afterword: what now? Climate-induced displacement after Copenhagen. In: McAdam J (ed) Climate change and displacement. Hart, Oxford, pp 239–246

Chazelnoel MT, Ionesco D (2018) A moment of opportunity to define the global governance of environmental migration: perspectives from the international organization for migration. In: McLeman R, Gemenne F (eds) Routledge handbook of environmental displacement and migration. Routledge, New York, pp 426–429

Collins AE (2013) Applications of the disaster risk reduction approach to migration influenced by environmental change. Environ Sci Policy 27:112–125. https://doi.org/10.1016/j.envsci.2012.10.005

Corry O (2012) Securitisation and 'riskification': second-order security and the politics of climate change. Millennium J Int Stud 40:235–258. https://doi.org/10.1177/0305829811419444

Dingwerth K, Kerwer D, Nölke A (eds) (2009) Die Organisierte Welt: Internationale Beziehungen und Organisationsforschung, 1. Aufl. Internationale Beziehungen, vol 12. Nomos, Baden-Baden

Dingwerth K, Lehmann I, Reichel E, Weise T, Witt A (2014) Many pipers, many tunes? Die Legitimationskommunikation internationaler Organisationen in komplexen Umwelten. In: Conceição-Heldt Ed, Koch M, Liese A (eds) Internationale Organisationen: Autonomie,

Politisierung, interorganisationale Beziehungen und Wandel. Nomos, Baden-Baden, pp 186–212
Eckhard S, Mattmüller J (2017) Verwaltungseinfluss und Verhandlungsergebnisse in internationalen Organisationen: Ein ressourcenbasiertes Modell. dms-der moderne staat-Zeitschrift für Public Policy, Recht und Management 10:229–250
Ege J (2016) Verwaltungsautonomie in internationalen Organisationen: Eine deskriptiv-vergleichende Analyse. Springer, Wiesbaden
Ege J, Bauer MW (2013) International bureaucracies from a public administration and international relations perspective. In: Reinalda B (ed) Routledge handbook of international organizations. Routledge, New York, pp 135–148
Elliott L (2010) Climate migration and climate migrants: what threat, whose security? In: McAdam J (ed) Climate change and displacement. Hart, Oxford, pp 175–190
Enkler J, Grohs S (2017) An den Grenzen der Lernfähigkeit? Lernkurven internationaler Verwaltungen nach der Finanzmarktkrise. dms-der moderne staat-Zeitschrift für. Public Policy, Recht und Management 10:251–270
Findlay AM (2011) Migrant destinations in an era of environmental change. Glob Environ Change 21:50–58. https://doi.org/10.1016/j.gloenvcha.2011.09.004
Fioretos O (2011) Historical institutionalism in international relations. Int Org 65:367–399
Flick U (2014) An introduction to qualitative research, 5th edn. Sage, London
Franck M (2018) UNHCR's perspectives and activities on displacement in the context of climate change. In: McLeman R, Gemenne F (eds) Routledge handbook of environmental displacement and migration. Routledge, New York, pp 403–407
George AL, Bennett A (2005) Case studies and theory development in the social sciences. MIT Press, Cambridge, MA
Gläser J, Laudel G (2004) Experteninterviews und qualitative Inhaltsanalyse als Instrumente rekonstruierender Untersuchungen. VS Verlag für Sozialwissenschaften, Wiesbaden
Guterres A (2007) Opening statement by Mr. António Guterres, UNHCR: Fifty-eighth session of the Executive Committee of the High Commissioner's Programme, Geneva
Gutner T (2005) Explaining the gaps between mandate and performance: agency theory and World Bank environmental reform. Global Environ Polit 5:10–37. https://doi.org/10.1162/1526380054127727
Haas EB (1990) When knowledge is power: three models of change in international organizations. University of California Press, Cambridge, MA
Hall N (2015) Money or mandate? Why international organizations engage with the climate change regime. Glob Environ Polit 15:79–97
Hall N (2016a) Displacement, development and climate change: international organizations moving beyond their mandates. Routledge, Abingdon
Hall N (2016b) The institutionalisation of climate change in global politics. In: Sosa-Nunez G, Atkins E (eds) Environment, climate change and international relations. E-International Relations Publishing, Bristol, pp 60–74
Hall N, Woods N, Kabra SS, Taranova Y, Kellerman M, Batten H (2015) Effective leadership in international organizations. World Economic Forum, Geneva
Hammerstad A (2014) The rise and decline of a global security actor: UNHCR, Refugee Protection, and Security. Oxford University Press, Oxford
Hanrieder T (2014a) Gradual change in international organisations: agency theory and historical institutionalism. Politics 34:324–333. https://doi.org/10.1111/1467-9256.12050
Hanrieder T (2014b) Local orders in international organisations: the World Health Organization's global programme on AIDS. J Int Relat Dev 17:220–241
Hantscher S (2016) The United Nations High Commissioner for Refugees' discourse on environmentally displaced persons: a double-edged sword? In: Rosenow-Williams K, Gemenne F (eds) Organizational perspectives on environmental migration. Routledge, London, pp 93–107
Hugo G (2010) Climate change-induced mobility and the existing migration regime in Asia and the Pacific. In: McAdam J (ed) Climate change and displacement. Hart, Oxford, pp 9–36

International Public Administration DFG Research Unit (2018) The emergence and development of administrative patterns and their effects on International Policy-Making (IPA). http://ipa-research.com/research-program/. Accessed 13 Mar 2019

IPCC (2012) Managing the risks of extreme events and disasters to advance climate change adaptation: a special report of Working Group I and II of the intergovernmental panel on climate change. https://www.ipcc.ch/pdf/special-reports/srex/SREX_Full_Report.pdf. Accessed 13 Mar 2019

Jackson NJ (2006) International organizations, security dichotomies and the trafficking of persons and narcotics in post-Soviet Central Asia: a critique of the securitization framework. Secur Dialogue 37:299–317

Jetschke A, Theiner P (2016) The diffusion of institutional design among regional organizations. https://papers.ssrn.com/sol3/papers.cfm?abstract_id=2735715. Accessed 13 Mar 2019

Kaiser R (2014) Qualitative Experteninterviews: Konzeptionelle Grundlagen und praktische Durchführung. Springer, Wiesbaden

Kälin W (2010) Conceptualizing climate-induced displacement. In: McAdam J (ed) Climate change and displacement. Hart, Oxford, pp 81–104

King G, Keohane RO, Verba S (1994) Designing social inquiry: scientific inference in qualitative research. Princeton University Press, Princeton

Kita SM, Raleigh C (2018) Environmental migration and international political security: rhetoric, reality and questions. In: McLeman R, Gemenne F (eds) Routledge handbook of environmental displacement and migration. Routledge, New York, pp 356–370

Kleine M (2013) Trading control: national fiefdoms in international organizations. Int Theory 5:321–346

Kolmannskog V (2012) Climate change, environmental displacement and international law. J Int Dev 24:1071–1081. https://doi.org/10.1002/jid.2888

Koremenos B, Lipson C, Snidal D (2001) The rational design of international institutions. Int Org 55:761–799

Kratochwil F, Ruggie JG (1986) A state of the art on an art of the state. Int Org 40:753–775

Krumm T (2009) Grundgedanken und Grundelemente qualitativer Forschung. In: Westle B (ed) Methoden der Politikwissenschaft. Nomos, Baden-Baden, pp 96–108

Kuckartz U (2009) Inhaltsanalyse. In: Westle B (ed) Methoden der Politikwissenschaft. Nomos, Baden-Baden, pp 334–344

Lehnert M, Miller B, Wonka A (2007) Na Und? Überlegungen zur theoretischen und gesellschaftlichen Relevanz in der Politikwissenschaft. In: Gschwend T, Schimmelfennig F (eds) Forschungsdesign in der Politikwissenschaft: Probleme - Strategien - Anwendungen. Campus Verlag, Frankfurt am Main, pp 39–59

Leuffen D (2007) Fallauswahl in der qualitativen Sozialforschung. In: Gschwend T, Schimmelfennig F (eds) Forschungsdesign in der Politikwissenschaft: Probleme - Strategien - Anwendungen. Campus Verlag, Frankfurt am Main, pp 204–222

Loescher G, Milner J (2011) UNHCR and the global governance of refugees. In: Betts A (ed) Global migration governance. Oxford University Press, New York, pp 189–209

March JG (1989) Footnotes to organizational change. In: March JG (ed) Decisions and organizations. Basil Blackwell, Cambridge, MA, pp 167–186

Martin LL, Simmons BA (2013) International organizations and institutions. In: Carlsnaes W, Risse T, Simmons BA (eds) Handbook of international relations, 2nd edn. Sage, London, pp 326–351

Martin S, Bergmann J, Wyss H, Rigaud KK (2018) Environmental change and human mobility: perspectives from the World Bank. In: McLeman R, Gemenne F (eds) Routledge handbook of environmental displacement and migration. Routledge, New York, pp 408–414

Mayring P (2014) Qualitative content analysis: theoretical foundation, basic procedures and software solution. SSOAR, Klagenfurt

McAdam J (2011) Environmental migration. In: Betts A (ed) Global migration governance. Oxford University Press, New York, pp 153–188

McAdam J (2017) From the Nansen initiative to the platform on disaster displacement: shaping international approaches to climate change, disasters and displacement. UNSW Law J 39:1518–1546

McLeman R, Gemenne F (2018a) Environmental migration research: evolution and current state of the science. In: McLeman R, Gemenne F (eds) Routledge handbook of environmental displacement and migration. Routledge, New York, pp 3–16

McLeman R, Gemenne F (eds) (2018b) Routledge handbook of environmental displacement and migration. Routledge, New York

Mearns R, Norton A (eds) (2010) Social dimensions of climate change: equity and vulnerability in a warming world. The World Bank, Washington, DC

Mehde V (2007) Die Entwicklung von Verwaltungswissenschaft und Verwaltungsreform in den letzten fünf Jahrzehnten. In: Ziekow J (ed) Entwicklungslinien der Verwaltungspolitik: Tagung aus Anlass des 50jährigen Bestehens der Deutschen Sektion des Internationalen Instituts für Verwaltungswissenschaften am 28. und 29. September 2006 in Speyer, 32nd edn. Nomos, Baden-Baden, pp 25–42

Meierding E (2013) Climate change and conflict: avoiding small talk about the weather. Int Stud Rev 15:185–203. https://doi.org/10.1111/misr.12030

Methmann C, Rothe D, Stephan B (eds) (2013) Interpretive approaches to global climate governance: (de)constructing the greenhouse. Routledge, New York

Myers N (1993) Environmental refugees in a globally warmed world. BioScience 43:752–761

Nielson DL, Tierney MJ, Weaver CE (2006) Bridging the rationalist–constructivist divide: re-engineering the culture of the World Bank. J Int Relat Dev 9:107–139. https://doi.org/10.1057/palgrave.jird.1800084

Nölke A (2000) Regieren in transnationalen Politiknetzwerken? Kritik postnationaler Governance-Konzepte aus der Perspektive einer transnationalen (Inter-) Organisationssoziologie. Zeitschrift für Internationale Beziehungen 7:331–358

Oels A (2012) From "securitization" of climate change to "climatization" of the security field: comparing three theoretical perspectives. In: Scheffran J, Brzoska M, Brauch HG, Link PM, Schilling J (eds) Climate change, human security and violent conflict: challenges for societal stability. Springer, Berlin, pp 185–205

Oels A, von Lucke F (2015) Gescheiterte Versicherheitlichung oder Sicherheit im Wandel: Hilft uns die Kopenhagener Schule beim Thema Klimawandel? Zeitschrift für Internationale Beziehungen 22:43–70

Pease K-KS (2016) International organizations, 5th edn. Routledge, New York

Platform on Disaster Displacement (2018) The protection agenda – key definitions. https://disasterdisplacement.org/the-platform/key-definitions. Accessed 13 Mar 2019

Portes A, Wind J de (eds) (2007) Rethinking migration: new theoretical and empirical perspectives. Berghahn Books, New York

Rittberger V, Zangl B (2003) Internationale Organisationen: Politik und Geschichte, 3rd edn. VS Verlag für Sozialwissenschaften, Wiesbaden

Rosenow-Williams K (2016) Environmental migration and the international red cross/red crescent movement. In: Rosenow-Williams K, Gemenne F (eds) Organizational perspectives on environmental migration. Routledge, London, pp 126–144

Rosenow-Williams K, Gemenne F (2016) Organizational perspectives on environmental migration. Routledge studies in development, mobilities and migration. Routledge, London

Rothe D (2012) Security as a weapon: how cataclysm discourses frame international climate negotiations. In: Scheffran J, Brzoska M, Brauch HG, Link PM, Schilling J (eds) Climate change, human security and violent conflict: challenges for societal stability, vol 8. Springer, New York, pp 243–258

Rothe D (2016) Securitizing global warming: a climate of complexity. Routledge, New York

Saha S (2012) Security implications of climate refugees in urban slums: a case study from Dhaka, Bangladesh. In: Scheffran J, Brzoska M, Brauch HG, Link PM, Schilling J (eds)

Climate change, human security and violent conflict: challenges for societal stability, vol 8. Springer, New York, pp 595–611
Scheffran J, Brzoska M, Brauch HG, Link PM, Schilling J (eds) (2012) Climate change, human security and violent conflict: challenges for societal stability, vol 8. Springer, New York
Schroeder MB (2014) Executive leadership in the study of international organization: a framework for analysis. Int Stud Rev 16:339–361. https://doi.org/10.1111/misr.12147
Seawright J, Gerring J (2008) Case selection techniques in case study research: a menu of qualitative and quantitative options. Polit Res Q 61:294–308
Siebenhüner B (2008) Learning in international organizations in global environmental governance. Global Environ Polit 8:92–116
Sosa-Nunez G, Atkins E (eds) (2016) Environment, climate change and international relations. E-International Relations Publishing, Bristol
Steffek J, Kortendiek N (2018) Participatory governance in international organizations. In: Heinelt H (ed) Handbook on participatory governance. Edward Elgar, Cheltenham, UK, pp 203–224
Subramanian N, Urpelainen J (2014) Addressing cross-border environmental displacement: when can international treaties help? Int Environ Agreements Polit Law Econ 14:25–46
Tallberg J, Sommerer T, Squatrito T, Jönsson C (2013) The opening up of international organizations: transnational access in global governance. Cambridge University Press, New York
Tänzler D, Ries F (2012) International climate change policies: the potential relevance of REDD+ for peace and stability. In: Scheffran J, Brzoska M, Brauch HG, Link PM, Schilling J (eds) Climate change, human security and violent conflict: challenges for societal stability, vol 8. Springer, New York, pp 695–705
The Nansen Initiative (2015) The Nansen initiative global consultation conference report, Geneva, 12–13 October 2015. The Nansen Initiative, Geneva. https://www.nanseninitiative.org/global-consultations/. Accessed 1 July 2019
Trombetta MJ (2014) Linking climate-induced migration and security within the EU: insights from the securitization debate. Crit Stud Secur 2:131–147. https://doi.org/10.1080/21624887.2014.923699
Türk V (1992) Das Flüchtlingshochkommissariat der Vereinten Nationen (UNHCR). Duncker & Humblot, Berlin
UNHCR (1951) Convention and protocol relating to the status of refugees. UNHCR, Geneva. https://www.unhcr.org/protection/basic/3b66c2aa10/convention-protocol-relating-status-refugees.html. Accessed 1 July 2019
UNHCR (2008) Climate change, natural disasters and human displacement: a UNHCR perspective. United Nations High Commissioner for Refugees, Geneva. https://www.refworld.org/docid/492bb6b92.html. Accessed 1 July 2019
UNHCR (2010) Convention and protocol relating to the status of refugees: with an introductory note by the office of the United Nations High Commissioner for Refugees. United Nations High Commissioner for Refugees, Geneva
UNHCR (2012) The state of the world's refugees 2012: in search of solidarity. The High Commissioner for Refugees, Geneva. https://www.unhcr.org/publications/sowr/4fc5ceca9/state-worlds-refugees-2012-search-solidarity.html. Accessed 1 July 2019
UNHCR (2017) Figures at a glance. United Nations High Commissioner for Refugees, Geneva. https://www.unhcr.org/ph/figures-at-a-glance. Accessed 1 July 2019
Verbeek B (1998) International organizations: the ugly duckling of international relations theory? In: Reinalda B, Verbeek B (eds) Autonomous policy making by international organizations. Routledge, New York, pp 11–26
von Lucke F, Wellmann Z, Diez T (2014) What's at Stake in securitising climate change? Towards a differentiated approach. Geopolitics 19:857–884. https://doi.org/10.1080/14650045.2014.913028
Warner K (2010) Global environmental change and migration: governance challenges. Glob Environ Change 20:402–413. https://doi.org/10.1016/j.gloenvcha.2009.12.001

Warner K, Hamza M, Oliver-Smith A, Renaud F, Julca A (2010) Climate change, environmental degradation and migration. Nat Hazards 55:689–715. https://doi.org/10.1007/s11069-009-9419-7

Weaver C, Leiteritz RJ (2005) "Our poverty is a world full of dreams:" reforming the World Bank. Global Gov 11:369–388

Wendt A (1992) Anarchy is what states make of it: the social construction of power politics. Int Org 46:391–425

Westle B (2009) Teilstrukturierte und offene Befragungsformen. In: Westle B (ed) Methoden der Politikwissenschaft. Nomos, Baden-Baden, pp 248–259

Zetter R (2010) Protecting people displaced by climate change: some conceptual challenges. In: McAdam J (ed) Climate change and displacement. Hart, Oxford, pp 131–150

# Chapter 2
# Environmental Changes and the Competing Perspectives on Environmentally Displaced Persons

This chapter introduces the case and thus presents a concise background of the academic approaches to the topics of environmental changes and disaster displacement. Under environmental changes, this thesis sums up both climate change-related changes, natural disasters that are non-related to climate change and environmental destruction that is non-related to climate change. In this context, the chapter outlines some of the leading security theories that serve as a background for many of the scholarly approaches to the topic. In doing so, it shows the development from a strictly environmental conflict debate up to the early 2000s toward the current—more nuanced—debate on climate change, natural disasters and the accompanying human implications (cf. Oels 2012, p. 185). In this way, the chapter outlines the security connotations with regard to environmental migration and displacement, and it presents a background on EDPs. Overall, this chapter lays the groundwork for the analysis.

## 2.1 Security Studies

As depicted above, the diverse impacts of environmental changes have received much attention in science, politics and the media. To broaden the security studies' perspective on EDPs, this section aims at clarifying the propositions that security studies make in general and on EDPs in particular. In other words, this chapter introduces security theory, first, to outline the linkage of security and environmental changes and, second, to better understand why and how environmental changes were and sometimes still are depicted as matters of international, national or human security by international actors such as the UNHCR.

The prominent approaches to security studies started to become more widespread in the 1990s, challenging traditional realist and neorealist theories (Jackson 2006, p. 300). In this context, the call for a broadening of the concept of security arose (Booth 1991, p. 317). The discussion about the meaning and conception of security



S. Hantscher, *The UNHCR and Disaster Displacement in the 21st Century*, Contributions to Political Science, https://doi.org/10.1007/978-3-030-19689-9_2

came about due to changes in the international environment after the Cold War (Rothe 2016, p. 84). That said, there has been a theoretically as well as practically oriented debate on the definition of security since the 1990s, which has extended far into the 2000s (Rothe 2016, p. 84). Scholars started to question the predominant military focus on security, arguing that this definition of security is too narrow. This restriction became apparent in the "pressures of the security dilemma, the growing appreciation of security interdependence, the widespread recognition that the arms race has produced higher levels of destructive power (. . .) and the realization of the heavy burden on economies of extravagant defence spending" (Booth 1991, p. 318). Scholars of security studies acknowledged that territorial threats and military security exist, but they nevertheless demanded the inclusion of other issue areas in the security agenda. Accordingly, scholars identified threats such as overpopulation, political oppression, economic collapse, ethnic rivalry, disease, terrorism, crime and the destruction of nature as new security challenges (Booth 1991). Additionally, the effects of 9/11 on international security perceptions have been immense. Scholars and politicians acknowledged that security challenges at the domestic level can have an international impact. Together with the national security-centered debate in, for example, the UN Security Council (UNSC) or think tanks, the concept of human security came to the fore. Here, the focus lies on "a shift of security policy to the single individual" (Rothe 2016, p. 84). In a nutshell, the changes at the international level led to changes in how security was perceived, and scholars ultimately addressed this development in security theory. Building on the environmental debates, the focus now shifts to how security theory has addressed these developments.

### 2.1.1 *Copenhagen School*

The securitization theory was originally developed by Buzan, Wæver and de Wilde. The Copenhagen School redefined security in a broader sense and developed elements to rethink security. From the perspective of the Copenhagen School, securitization describes an extreme form of politicization, because a topic is "dramatized and presented as an issue of supreme priority, by labelling it as security" (Buzan et al. 1998, p. 26). Securitization is understood as "the move that takes politics beyond the established rules of the game and frames the issue either as a special kind of politics or as above politics" (Buzan et al. 1998, p. 23). In case an issue is taken off the security agenda and shifted back to the normal political realm, the Copenhagen School refers to the issue as being desecuritized (Jackson 2006, p. 301). The scholars of the Copenhagen School have thus positioned themselves in favor of a politics of deliberation and against securitization (Corry 2012, p. 236).

Central to securitization theory are so-called speech acts (Buzan et al. 1998, p. 26). In this process, securitization is "constituted by the intersubjective establishment of an existential threat with a saliency sufficient to have substantial political effects" (Buzan et al. 1998, p. 25). The language of security is used by the securitizing actor in order to articulate an already politicized issue as an existential threat to a referent object. According to the Copenhagen School, NGOs, IOs or states can fulfill the role of a securitizing actor (Jackson 2006, p. 301). Most importantly, to speak of successful securitization, the audience of the speech act first must accept the securitized issue as an existential threat, and second, must tolerate "violations of rules that would otherwise have to be obeyed" (Buzan et al. 1998, p. 25). Due to the analytical vagueness of the precise role of the audience in the securitization theory, some scholars have proposed re-conceptualizing the role of the audience in the process (Léonard and Kaunert 2011) or focusing on the exact meaning of the speech act in securitization (cf. Balzacq 2005; Huysmans 2011). As an example, Rothe refined securitization as "an ongoing discursive struggle" (Rothe 2016, p. 31).[1]

A successful securitization of EDPs implies that an actor has framed the environmentally displaced in national or international security terms, e.g., as threats to national sovereignty or international peace and security; the speech acts need to include traditional security responses, such as the involvement of the military; and the protection of EDPs needs to allow for extraordinary measures that are normally perceived as intolerable by the public. As an example, the securitization of the environment might "legitimize a political state of exception, in which drastic mitigation measures are adopted using undemocratic procedures" (Oels 2012, p. 185) or involve "limiting national sovereignty for the sake of climate protection, restricting individual liberties through rationing carbon emissions, or opposing existing ethical rules e.g., by legitimizing geo-engineering" (Methmann et al. 2013, p. 119).

For the case of EDPs, the Copenhagen School's views need to be read as a contextualized securitization (cf. von Lucke et al. 2014, p. 860). As such, the more recent definitions of securitization have relaxed the strict focus on grammar and speech acts, i.e., the articulation of threats that have not arrived at an existential level yet are nevertheless considered (von Lucke et al. 2014). This broadening of the definition has been motivated by earlier research on the securitization of climate change (Methmann et al. 2013, pp. 91–92; Trombetta 2014, p. 136), which calls for a context-specific approach and specific awareness when it comes to securitization processes in the environmental sector. Although the depiction of environmental changes in politics, science, and the media peaked in 2007 in terms of security references, most scholars concluded that there was and still is no securitization of environmental changes in the strict sense of the Copenhagen School: Reality shows that states have not pushed for extraordinary and undemocratic measures (among others, Oels 2012, p. 185; Oels and von Lucke 2015). For more details on the debate concerning the securitization of environmental changes, see as proponents, Brauch (2012) and Brzoska (2009), and as a critic, Trombetta (2008).

Overall, the securitization theory helps to identify those who are threatened and uncovers the actors who implement the securitizing move. Moreover, the theory allows an examination of how a securitization process unfolds by shifting the focus to discursive speech acts. Looking at the theory from a more critical stance, some scholars have pointed out that securitization theory has a predominantly European

[1]For an overview of the critique on the speech act in the securitization approach, see Rothe (2016).

perspective because it solely reflects European security concerns (Jackson 2006). Also, the theory lacks a gender-sensitive perspective and ignores all other political communication besides speech-acts (Jackson 2006, p. 300). Scholars have criticized the two schools for lacking clear methodological detail and for being too imprecise to clearly point to securitization when it is present (Corry 2012, p. 240). Lastly, the role that is given to the state in the framework is quite prominent; here, scholars have perceived the role of the state as the securitizing actor as being overly central (Jackson 2006, p. 301).

### 2.1.2 *Paris School*

Some scholars have interpreted the Copenhagen School and the Paris School as competing standpoints with opposite perceptions of security, and they have therefore seen a battleground between the two theories (Bourbeau 2014, p. 188). Contrary to this, the thesis at hand is in line with research by Trombetta (2008) in interpreting the Paris School as an insightful further development of Buzan, Wæver, and de Wilde's framework. The sociological approach of the Paris School essentially extends the discourse argument of the Copenhagen School to non-discursive practices. Additionally, the framework takes the institutions into account that embed the speaker of a speech act into his or her surroundings, and thus aims at unveiling the power positions and institutional struggles behind a securitizing act.

According to Bigo and Balzacq, the central contributors to the Paris School framework, "it is possible to securitize certain problems without speech or discourse and the military and the police have known that for a long time. The practical work, discipline, and expertise are as important as all forms of discourse" (Bigo 2000, p. 194). As a continuation of the Copenhagen School, scholars added the role of practice to the existing tools of securitization. Balzacq thus pointed out that "because securitization is the product of such a complex repertoire of causes, an investigation of a unique factor (e.g., rules of speech acts) may fail if other elements exert a significant influence on the process" (Balzacq 2005, p. 193). The Paris School's understanding of security helps to tackle the issue of security professionals and security agencies that authoritatively claim to "identify threats [and] provoke existing fears to promote their own institutional interests" (Trombetta 2014, p. 137). Here, a securitization act is not dependent on the presence of exceptional measures; rather, the normal governance mechanisms will be raised to the traditional security field with, for example, policing mechanisms or surveillance (Trombetta 2014). In this understanding of security, technical and professional expertise becomes significant, and scholars can focus on the degrees of institutionalization that have an impact on the securitization process. Following this line of thought, the "spectacular speech acts and calling for new or more extensive forms of governance are likely to occur when institutions are limited as in the case of environmental migration" (Trombetta 2014, p. 138). Especially relevant is the Paris School's focus on the structural position of the speaker of the speech act; here, securitization is the result of power positions. Analyzing a speech act with this analytical framework thus allows better insight to be gained into the institutions behind the person who securitizes an issue (Bigo 2002, p. 74).

In the context of environmental changes, the Paris School assumes a 'climatization' of the security field, i.e. professionals of security paint a picture of environmental changes as threats to their daily business (Oels 2012, p. 185). Examples of this climatization are the military, the police or the defense ministries that have practically included environmental changes in their early warning systems and planning studies (Oels 2012, p. 185). Building on how the Copenhagen School and the Paris School look at the matter of environmental migration and displacement, it shows that the people of concern—the people who must migrate due to environmental changes—are not sufficiently depicted in these two schools. In other words, the security theories mentioned above do not provide sufficient theoretical background for the empirical puzzle of this thesis. To be able to fully explain the UNHCR's depiction of EDPs, it is necessary to introduce the concept of Human Security in the section below.

### *2.1.3 Human Security*

The end of the Cold War signified a shift in the understanding of security, as new threats appeared on the international scene. Due to the emergence of powerful non-state actors, state-centric approaches reached their limits, and the traditional concept of power began to be contested (Tadjbakhsh and Chenoy 2007, p. 1). The Human Security concept goes back to a definition that can be found in a 1994 report of the United Nations Development Programme (UNDP), the so-called Human Development Report. This report defined human security rather broadly as the freedom from want and the freedom from fear, and thus incorporated all kinds of threats that undermine human well-being (Webersik 2010, p. xviii). It claimed that the traditional concept of security, because of its state-centeredness, is too narrow and neglects the individual's security (Kaldor 2007, p. 182). The Human Development Report identified "seven core elements that together make up the concept of Human Security—economic security, food security, health security, environmental security, personal security, community security and political security" (Kaldor 2007, p. 182). According to this concept, the term 'security' should go beyond states and should include significant threats to individuals such as hunger, poverty, diseases, violence, political and gender repression, etc. The concept can be described as a host of preventive measures to prevent conflicts and disputes from arising and escalating (Schlesinger 2008, pp. 16–18). Human security shifts the focus from the state to the individual and includes not only the actual absence of physical violence but also the well-being and dignity of human beings (Tadjbakhsh and Chenoy 2007, p. 1). Yet, the developments since 9/11 have demonstrated a turning away from human security and a stronger focus on the national security of the countries in the global North. In other words, if the threats to human security in the global South do not impact the national security concerns of the global North, human security is not a strategic ground on which states act (Oels 2012, p. 185).

Nevertheless, scholars have recently re-interpreted the Human Security approach and have applied it to the contexts of the environment, gender and refugees (among others, see Rosenow-Williams and Behmer 2015, p. 188). Applied to the topic of environmental changes, human security would imply a shift in the policy agenda toward sustainable development (Oels 2012, p. 185). Also, a Human Security concept of EDPs would imply a shift in awareness toward those individuals who have suffered as a result of environmental changes and the way they have tried to cope with—sudden or long-term—changes in the environment.

In general, the critique of the Human Security concept is as follows: Even though the concept can fill some gaps due to its focus on the individual, the concept has a theoretical vagueness that makes it difficult to apply empirically in a comparable manner. Also, the concept is undeniably normative, which has resonated in some of the scholarly work. Lastly, as of 2018, it appears that the concept, when compared to the securitization perspective of the two schools, is not as congruent with the security perceptions of most states and international actors. Consequently, scholars who have aimed at explaining security conceptions since the early 2000s have usually disregarded the Human Security concept.

As outlined above, this section answers the 'what' question; in other words, what are we dealing with? Therefore, the next section introduces environmental changes by giving background information on the terminologies of climate change and natural disasters. Additionally, it outlines the extent to which scholars and practitioners have perceived these phenomena from a security perspective. Therefore, a concise background on the two most prominent interpretations, environmental security and environmental conflict, is given. The aim is to familiarize the setting of this analysis and to be able to classify the chosen case study into the broader topic of environmental changes and security.

## 2.2 Environmental Changes

This thesis uses the term 'environmental change' as an umbrella term for climate change and natural disasters. When talking about environmental and climatic changes, definitions of climate and climate change are needed. As an example, Butler and Gates, who have focused on the impact of the climate on African range wars, quoted the World Meteorological Organization, defining climate as "the statistical description (mean and variability) of surface weather conditions such as temperature, precipitation, and wind over a period of 30 years" (Butler and Gates 2012, p. 26). Similar to this understanding of climate is the Intergovernmental Panel on Climate Change's (IPCC) definition of climate change: "A statistically significant and persistent (decades or longer) variation in the mean state or variability of climate" (IPCC 2001, p. 1). The IPCC specifically stressed that climate change may result from natural and internal processes, as well as external forcing.

Additionally, climate change may be rooted in ongoing anthropogenic changes in land use or in the composition of the atmosphere (IPCC 2001, p. 1). Furthermore, Scheffran et al. differentiated between climate change as the "natural variability in the climate system, which has fluctuated between warm and cold periods during the Holocene since the end of the glacial periods 12,000 years ago" (Scheffran et al. 2012a, p. 3), and the global warming during the Anthropocene. This period of earth history stands for the era since the Industrial Revolution, in which production as well as consumption patterns have directly impacted the earth's system. Since the 1970s, the Anthropocene has increasingly become an object of scientific research (Scheffran et al. 2012a, p. 3).

The problematic structure of climate change is well known and can be summarized as follows: Humans emit greenhouse gases (GHGs)—particularly carbon dioxide ($CO_2$), but also methane, nitrous oxide, and hydrofluorocarbons (HFCs) through consumption and production. These emission-flows accumulate into stocks of GHGs in the atmosphere, and "the rate of accumulation depends upon the earth's carbon cycle whereby $CO_2$ is reabsorbed into the oceans and land" (Helm and Hepburn 2009, p. 36). Over time, the accumulated GHGs trap heat, and the result is global warming. As the planet warms, the climatic conditions change, which directly affects human and animal life through rising sea levels and events such as storms, floods, and droughts (Helm and Hepburn 2009, p. 36). The basic science of global warming is more or less well known and understood, whereas the causes and especially the impacts of climate change are highly complex.[2] The scientific status quo on climate change can be understood as the result of two preceding debates: The first debate focused on whether the climate is changing. As a consensus evolved that the earth is getting warmer, the controversy shifted to a second debate about whether human beings are responsible for climate change. The third—and current—debate therefore involves questions concerning the necessary actions of states and other actors in addressing the consequences of global warming (Gartzke 2012, p. 178).

Whereas climate change has been defined above, the term 'natural disaster' is often specified as fast-onset environmental events,[3] in other words, a "relatively large, non-linear change in an environmental variable, over a relatively short interval of time, which can have a significant impact on people" (Lenton 2013, p. 61). Examples of these extreme events are tsunamis, earthquakes, floods, volcanoes, cyclones, tidal waves, landslides, and hurricanes (Collins 2013, p. 114). As these events typically attract a lot of national and international attention, these rapid-onset[4] environmental changes usually trigger humanitarian reactions and emergency aid (Collins 2013, p. 114). The extent to which people feel the impact of these events

[2]For an overview of climate impacts, see Reusswig and Lass (2013).

[3]For a categorization of environmental shocks, see Lenton (2013, p. 61), who differentiated between extreme events such as large earthquakes and tsunamis, abrupt swings, e.g. an El Niño event, and a tipping point where a system abruptly leaves its current state.

[4]Contrary to these rapid-onset environmental changes, research has also been conducted on slow-onset changes such as e.g. droughts, failing harvests, and rising sea-levels (Collins 2013, p. 114).

depends both on the question of vulnerability (Schilling et al. 2012) and on the exact nature of the environmental event (Lenton 2013, p. 61). The IPCC defined vulnerability[5] as the propensity or predisposition to be adversely affected, and it also mentioned the term in its definition of 'disaster': "Severe alterations in the normal functioning of a community or a society due to hazardous physical events interacting with vulnerable social conditions, leading to widespread adverse human, material, economic, or environmental effects that require immediate emergency response to satisfy critical human needs and that may require external support for recovery" (IPCC 2012, p. 17).

Predictions of the occurrence of natural disasters are controversial and present varying results for the next 10–50 years.[6] Nevertheless, there are data that show an increase in the frequency of heavy precipitation events at the global scale, although these are inconsistent between different models. The IPCC argued that the global average frequency of tropical cyclones will fall, whereas the intensity of these kinds of cyclones is likely to increase on average to the end of the century (Black et al. 2013b, p. 34). Further, due to higher sea levels, the occurrence of coastal flooding will likely increase (Black et al. 2013b, p. 34). Although the 'IPCC Special Report on Extreme Events' of 2012 was able to point out the frequency and geographical distribution of extreme weather events, a "robust regional monitoring or modelling of change" (Black et al. 2013a, p. 1) is still lacking. However, such a warning model would be helpful to pin down the future changes in certain regions. According to the 'IPCC Special Report', the meager available data give a comparable insight into extreme events, and they mostly concern changes in the frequency and intensity of extreme events (IPCC 2012, p. 20). More specifically, the future trends on a global scale are sometimes more reliable (e.g. temperature extremes) or less reliable (e.g. droughts) than some of the regional trends (IPCC 2012, p. 20). As an example, the International Disaster Database EM-DAT shows an increase in the frequency of natural disasters that affect people, and the data especially highlight the floods and storms that have this impact (Black et al. 2013b, pp. 33–34). Explanations for this trend range from the growing populations who live within disaster zones, as well as an increase in the insured property and economic assets that can be found in these areas (Black et al. 2013b, pp. 33–34).

Building on the terminology introduced above, the focus now shifts to how these environmental changes have been perceived by scholars so far. The reference to security implications is very common, and the leading interpretations of environmental changes are in terms of 'environmental security' and 'environmental conflict'. The following two sections summarize both interpretations and thereby assign the case study to the general topic of environmental changes.

[5]For a differentiation between risk and vulnerability, see Alexander (1997, p. 291). Alexander understands risk as the product of hazard and vulnerability.

[6]For more details on the prediction of natural hazards from satellite imagery, see Gillespie et al. (2007).

### 2.2.1 Environmental Security

Concerning the approaches to environmental changes and security, one of the main arguments of this debate implies that "instability elsewhere directly impacts security in Western states" (Rothe 2013, p. 115). The diversified literature on security and environmental changes[7] contains two main motives: on the one hand, the normative responsibility and, on the other hand, the national interest to provide assistance. A third main debate concerns the development-security nexus that especially focuses on political concepts "such as vulnerability, human security and resilience" (Rothe 2013, p. 115). Following this line of thought, the large-scale displacement of people due to environmental changes likely undermines their respective development (Barnett and Webber 2010, p. 40). Overall, the increasing attention to security considerations in the policy field of environmental changes is a relatively new development (Kurtz 2012, p. 669). Amongst others, Rothe (2016) shed light on the environmental security discourse and concluded that the alertness to climate change in 2007—with the Security Council debates on climate change and the IPCC's Nobel Peace Prize—started much earlier. In the 1990s, the debate that linked environmental and security issues was an addition to the preexisting debate on a broadened security agenda. In other words, the climate change and security discourse grew out of the broader environmental security discourse (Rothe 2013, p. 109). During this first episode of environmental security research, climate change was identified by many authors as a potential source of instability (Meierding 2013, p. 188). Homer-Dixon even compared climate stress to the arms race of the Cold War and the proliferation of nuclear weapons among rogue states, thus pointing to the challenge it poses to international security (Homer-Dixon 1994). In the 2000s, the causes of terrorism and its socio-economic roots attracted a great deal of attention, which led to a growing securitization of matters of development. These kinds of risk and security assessments are apparent in numerous articles between 2007 and 2008 (Meierding 2013, p. 188). Research on environmental changes and security usually uses the variables of environmental degradation, food security, agricultural productivity, population growth, societal (in-)stability and weather conditions, such as temperature, precipitation and droughts (Butler and Gates 2012, p. 26; Schilling et al. 2012, p. 12). Recent approaches have therefore integrated the different research fields of climate science, conflict research, social geography and environmental science (Schilling et al. 2012, p. 12). Concerning the geographical focus, the majority of researchers have used case studies on the MENA region, sub-Sahara Africa, South-East Asia and South-Asia (Bangladesh, India, Pakistan), and the Pacific Islands (Schilling et al. 2012, p. 12).

Political leaders have also depicted environmental changes—especially climate change—as a security threat: President George W. Bush's statement in 2002—that

[7]Cf. Baldwin and Bettini (2017); Boas and Rothe (2016); Corry (2012); Kita and Raleigh (2018); von Lucke et al. (2014); Meierding (2013); Methmann et al. (2013); Oels and von Lucke (2015); Rothe (2012, 2016); Scheffran et al. (2012a); Trombetta (2014).

the world has to fight against poverty because hope is an answer to terror—is just one example of how poverty-related grievances became inseparable from the security of the global North (Rothe 2013, p. 114). The UN Secretary General Ban Ki-Moon asserted numerous times that environmental changes can have negative human security implications, and President Barack Obama made clear that the current dangers to national security are interlinked with the long-term threat of climate change, which might also be related to violent conflict (Meierding 2013, p. 185). By this means, the fight against poverty has been interpreted as a national interest. Also, the concern over so-called failed states gained momentum and contributed to the manifestation of the development-security nexus in the Western security debates. Failed states might threaten neighboring states with refugees and political instability, and they therefore have international security consequences (Rothe 2013, p. 114). In 2004, the UN Secretary-General's High-level Panel on Threats, Challenges and Change stated in a report that the "biggest security threats we face now, and in the decades ahead, go far beyond states waging aggressive war. They extend to poverty, infectious disease and environmental degradation; war and violence within states; the spread and possible use of nuclear, radiological, chemical and biological weapons, terrorism; and transnational organized crime" (UN Secretary-General 2004, p. 1).

The debate in the UN after 2007 on the possible security implications of environmental changes provides an example of the strategic use of security references as part of the legitimation in the process of securitization (Kurtz 2012, p. 669). As representatives of non-governmental actors that have functioned as drivers of the debate, one can name the British International Alert (IA) or the American Center for Strategic and International Studies (CSIS) (Rothe 2012, p. 249). In 2009, the then UN Secretary General introduced the different channels through which climate change can affect security (UN General Assembly 2009): The first channel, vulnerability, implies that climate change has a negative effect on food security and human health, and thereby increases human exposure to extreme events. The second channel is development; climate change can slow down development processes and undermine the capacity of states to maintain stability. As a third channel, the concepts of coping and security are mentioned, thus focusing on migration and competition over natural resources, as well as the coping responses of communities. Ultimately, climate change might increase domestic conflicts and lead to international repercussions. Statelessness acts as the fourth channel; the disappearance of territory has implications for questions of rights, security and sovereignty. International conflict is the fifth channel and implies that the effects of climate change have an impact on international cooperation, especially concerning shared international resources.

An argument for the interlinkage of environmental changes and security aspects has been made in several reports of governmental and non-governmental institutions; the possible consequences are migration, conflicts over scarce resources, and the threat to the survival of small island states (Kurtz 2012, p. 669). The actors who have involved themselves in the debate on security politics and environmental changes are government officials, for example from Germany (the WBGU–German

Advisory Council on Global Change), Great Britain (the National Security Strategy of the United Kingdom of 2008) and the USA (e.g., the US Government, the White House, the National Security Strategy of the USA of March 2006) as well as IOs, such as the NATO (Rothe 2012, p. 249). As an example, the Ministry of Defense of the UK entered a contract with the UK Meteorological Office to perform research that focuses on the climate change and conflict linkage (Elliott 2010, p. 180). Additionally, the IPCC and former US Vice-President Al Gore received the Nobel Peace Prize for their efforts to build knowledge on climate change. In other words, "governments, international organizations and NGOs have increasingly directed their attention to climate change as a security issue and a likely source of conflict" (Elliott 2010, p. 179).

### 2.2.2 Environmental Conflict

Although the linkages between environmental changes and security aspects are quite commonly acknowledged, the link between environmental degradation and violent conflict has received similar attention by researchers. In the 1990s, the focus of conflict researchers was on the likelihood of violent conflict in developing countries due to environmental degradation (Rothe 2013, p. 109). Here, the focus was on the causal link that leads from resource scarcity to economic inequalities and culminates in the likeliness of ethnic clashes (Saha 2012, p. 599). Homer-Dixon, who led the so-called Toronto group, is one of the most well-known representatives of this research. By using qualitative case studies, the group tested arguments originating in neo-Malthusianism and applied them to resource scarcity, population growth and the possibility of violent conflict (Homer-Dixon 1991, 1994). The group chose different kinds of environmental changes—such as water degradation, the reduction of the stratospheric ozone layer, deforestation and the depletion of global fish stocks and climate change—and looked at the effects they have on violent conflict. The results are as follows: None of the changes mentioned above single-handedly explain the occurrence of conflict, but a combination of factors could indeed cause conflict and instability (Homer-Dixon 1994). The background of this argumentation lies in conflict studies: One assumes that the scarcity due to environmental changes leads to a worsening of the living conditions of the population. As a consequence, the population becomes more vulnerable to infectious diseases, and more people join dissident activities because the opportunity costs of doing so are lower than before (Saha 2012, p. 599). This research strand also takes the respective local triggering factors into consideration, because environmental degradation, resource mismanagement and economic stagnation are all causes of communal violence. Therefore, the central question is "which immediate and contextual factors are able to account for the spatial and temporal dynamics in how local resource competition escalates to the use of armed force?" (Saha 2012, p. 599)

This strand of research also constructs an indirect link between environmental degradation and instability. An example is the extensive logging in tropical areas that desalinates the soil and contributes to erosion. As a consequence, the land becomes much more vulnerable to natural hazards such as flooding and landslides. The decreased land productivity and the eradication of natural defenses might ultimately lead to poverty and migration, Haiti being one example of this development (Saha 2012, p. 599). Nevertheless, this environmental insecurity does not afflict all countries similarly, but varies across space and time (Saha 2012, p. 599). Overall, there is a rising scholarly interest in the consequences of environmental changes, such as the possible scarcity of resources and the rise of sea levels, as well as the possible social impacts, such as political violence and conflict (Gartzke 2012, p. 177). The assumption that communities and states might fight over water, soil and oil as a consequence of environmental change spread quickly into the policy-making cycles, triggered by researchers such as Homer-Dixon (1994). Scholars presented statistical data on the worsening of civil war in developing countries, which they traced back to recent environmental changes (Gartzke 2012, p. 177), following the logic that the scarcity of resources leads to an increase in conflicts. Generally speaking, the environmental conflict argumentation takes the aspect of *change* in environmental changes seriously: First, change leads to uncertainty on the question of whether resources are available, and second, changes can occur over a longer period of time, e.g. in (inter-) national power relations (Gartzke 2012, p. 180). Adding to this, researchers have focused on vulnerable groups and have investigated the increase of conflict emanating from those who occupy an inferior role in society (among others, see Meierding 2013, p. 194; Scheffran et al. 2012b). For this reason, vulnerable groups have a higher chance of becoming victims of environmental changes, and an aggressive response might be the only way of voicing their disapproval of environmental policies. Some studies have even linked the destabilizing effect of environmental changes with radicalization and terrorism. South-Asia, especially Bangladesh, and Southeast-Asia are representatives of the "combination of deteriorating socio-economic conditions, radical Islamist political groups, and environmental insecurity brought on by climate change" (Saha 2012, p. 601). The argument goes that the increasing number of weak and failing states serve as global terrorism bases, as exemplified by Somalia (Lennon and Smith 2007, p. 1). Additionally, some studies have mentioned EDPs in this context and have argued that tensions between the immigrant community and the target country population might trigger radical tendencies or even full-fledged radicalization (Lennon and Smith 2007, p. 1).

Nevertheless, the research has shown no definite empirical support for the environmental conflict argument (Meierding 2013, p. 194): The resources that become scarce due to environmental changes are already scarce (or unavailable) in certain regions, and "these regions are often more peaceful than the places where such resources are abundant" (Gartzke 2012, p. 180). Europe, for example, depends on others for fuel, but is peaceful nonetheless and thus exemplifies that scarcity by itself is not a sufficient indicator for conflict. And even though some agricultural producers might suffer from environmental impacts, they do not necessarily possess the martial strength to impose their will (Gartzke 2012, p. 180). In addition, some environmental conflict sceptics from peace and security studies have pointed to the missing empirical data for 'climate wars' (Nordas and Gleditsch 2007).

Environmental conflict is just one explanation out of many, because multiple factors affect this chain, and conflicts are seldom caused solely by environmental changes (Tänzler and Ries 2012, p. 696). Moreover, some long-term macro statistical studies have found an increase in conflicts due to a colder instead of a warmer climate, and since the mid-twentieth century—the period in which global warming accelerated—interstate conflicts have declined (Gartzke 2012, p. 178). Hence, macro studies have shown that violent conflicts between or within European countries or (civil) war in China due to global warming is not likely, and the burning of fossil fuels rather builds "economic and political systems that appear less inclined to certain forms of violent conflict" (Gartzke 2012, p. 178). Summing up, the media and politicians have misinterpreted or even falsely represented most of the research results—which led to global warming being associated with a rise in conflicts (Gartzke 2012, p. 178).

## 2.3 Environmentally Displaced Persons

Before going into more depth concerning the topic of environmentally-induced displacement and migration, an initial discussion of the use of the term 'refugee' is necessary. Political actors mostly employ the term 'climate refugee' to achieve their respective strategic goals. Whereas some use the term to call attention to future climate change scenarios or to advocate against environmental degradation, others utilize the refugee connotation to push for stricter migration laws (Methmann et al. 2013, p. 95). Concerning the distinction between refugee and displaced person, the term 'refugee' refers to those people who are eligible for a political and legal status, and who have crossed a national border. Refugees therefore enjoy "protection, aid or collective reception in camps" (Meertens and Segura-Escobar 1996, p. 166) in contrast to people who are forcibly displaced. States deal with the latter in an individual manner, and the political discussion about their protection remains controversial, as they "fall within a vacuum of responsibility within their own countries because of a rift between the affected population and the government or an opposing force" (Meertens and Segura-Escobar 1996, p. 167).

The literature on environmental displacement and migration has developed from a one-dimensional debate over the apparently growing number of EDPs (e.g. Myers 2005) toward a multi-faceted discussion concerning multiple drivers of migration, with environmental changes being just one driver among others (e.g. Black et al. 2013b; Piguet 2008). Parallel to these debates, scholars currently do not aim to pin down the exact number of people affected by disasters, but rather look at where migration systems are already in place and what the possible future migration paths will be (Findlay 2011, p. 50). For the most part, the discussion of how to define EDPs and whether a new category of displaced persons is an advantage or disadvantage is still ongoing.

### 2.3.1 *EDPs from the Perspective of Migration Studies*

Initially, it is important to note that geographers have acknowledged the environment as one trigger of migration in their attempt to establish a better understanding of migration theory. Nevertheless, throughout the twentieth century, the factor of the environment vanished in migration research (Piguet 2013, p. 150). The underlying explanations of why the environment disappeared in migration studies are, first, the predominantly Western idea that the environment does not have a large impact on progress; second, that migration studies increasingly focused on the economic motivations for migration; and third, that the research strand of refugee studies predominantly holds a political paradigm that is based on the 1951 Refugee Convention and which thus excludes environmental push factors (Piguet 2013, pp. 151–152).

Overall, most researchers have agreed that "environmental change is not a straightforward 'migration trigger'. It is very difficult to disentangle the effects of environmental change from other migration drivers" (Geddes and Somerville 2012, p. 1016). Therefore, scholars have stressed the need for tools to manage climate risks generally (among others, Tänzler and Ries 2012, pp. 695–696). As one example, the question arises of whether existing efforts in e.g. peacebuilding activities should be integrated into efforts to stabilize the situation of a disaster-prone country. Tänzler and Ries argued that so far, "little attention has been paid to substantiate the impacts of climate change and climate change policies on conflict-prone countries" (Tänzler and Ries 2012, pp. 695–696). Possible unintended side-effects of mitigation and adaption measures on the overall situation of a (post)conflict constellation need to be taken into consideration to develop "conflict-sensitive climate politics" (Tänzler and Ries 2012, p. 703). Schilling et al. thus called for more research on pre-emptive policies, particularly agricultural and socio-economic policies (Schilling et al. 2012, pp. 21–22). They argued that it is essential to explore the extent to which these policies focus on societal implications and on environmental change.

Even though it is difficult to clearly identify the environment as one migration trigger, by the end of the 1980s, the topic of the environment reemerged in migration studies. This was due to publications by the IPCC, the Worldwatch Institute and the United Nations Environment Programme (UNEP) that triggered subsequent debates on 'environmental refugees' and large migratory streams (Piguet 2013, p. 153). When looking at the current academic literature on environmental extremes and disasters, it is interesting to note that these publications have underemphasized mobility in general and migration in particular as adaption mechanisms (Black et al. 2013a, p. 1). Black et al. stressed the importance of migration in "societal and economic trajectories, in urbanization and in cultural exchange" (Black et al. 2013a, p. 1). Still, most of the literature has only focused on the displacement dimension of migration in relation to environmental disasters. Scholars have thus perceived migration as a problematic reaction to environmental change. Contrary to this, Black et al. argued that migration is just *one* kind of human adjustment to environmental changes (Black et al. 2013a, p. 2). That said, migration due to environmental changes is a very complex concept—the changes in the environment are just one factor among many (Barnett and Webber 2010, p. 37). When looking at migration outside of the extraordinary impacts of disasters, it becomes clear that migration is a social phenomenon and a long established "system of social and demographic interaction and change" (Black et al. 2011, p. 4), and overall, a human process (Warner et al. 2010). This human process implies that a large proportion of the population prefers to stay instead of moving or migrating—even when they are exposed to extreme events such as the Japanese tsunami of 2011 (Black et al. 2011, p. 6). According to Barnett and Webber, "migration may be an effect of environmental change (an outcome that migrants might rather have avoided) as well as an adaptation response (to avoid or adjust to an even more undesirable outcome)" (Barnett and Webber 2010, p. 38).

Additionally, changes in ecosystem services also trigger migration because they directly link the environment to well-being.[8] Following this line of thought, environmental changes also have an indirect effect on migration because they have an impact on economic and political drivers. The above-mentioned research results are important when working on slow-onset changes in the environment, as well as when focusing on the long-term effects of sudden-onset disasters. Concerning slow-onset environmental changes, extensive research exists on the possible drivers of migration (Black et al. 2011, p. 3).[9] As an example, the triggers that affect the decision to migrate are "economic, political, social, demographic and environmental" (Black et al. 2011, p. 3), and they often occur simultaneously. For example, some case studies have focused on fast-onset environmental changes such as Hurricane Katrina. This case exemplified that the return of the displaced people depends heavily on factors such as age, income and severity of damage; thus, the old and poor from the extremely damaged neighborhoods were the slowest to return (Black et al. 2011, p. 7). Other case studies have included Mozambique in the early 2000s, which showed the extent to which people resettle after flooding, and Vietnam, which represented a case in which flooding ultimately led to relocation of the people of concern (Warner 2010, pp. 405–406). Furthermore, the 'IPCC Special Report' specified those risks that lead to displacement, e.g. landlessness, joblessness, homelessness, marginalization, food insecurity, increased morbidity, loss of access to common property resources, and social disarticulation (IPCC 2012, p. 80). Moreover, research done by Findlay (2011) and Warner et al. (2010) pointed to social system characteristics and encouraged analyses of whether and where people move. Overall, it is necessary to look at existing migration systems that deliver insight into the decisions, argumentations and choice of places, and to acknowledge that they are all socially constructed (Findlay 2011, p. 53). In other words, environmental factors are "multiplier effect[s], contributing to existing migration regimes" (Findlay 2011, p. 53).

---

[8]For more information on ecosystem services, see Black et al. (2013a, p. 2).

[9]Examples for slow-onset environmental crises are deforestation, drought, disease outbreak and cumulative pollution (Black et al. 2013b, p. 33, Collins 2013, p. 114).

### 2.3.2 EDPs from the Perspective of Disaster Displacement Studies

Disaster studies have built on the paradigms of, first, natural hazards and, second, the construction of disaster causality (Collins 2013, p. 115). Whereas some authors have focused on human vulnerability, others have taken the resilience to disasters as their central research focus (among others, see Schilling et al. 2012). Overall, disaster and displacement studies have combined elements from social, economic and environmental analyses to be able to conduct disaster vulnerability analyses or models of displacement (Collins 2013, p. 115; Kniveton et al. 2011, p. 34). In this context, the people who are dependent on natural resources and who are poor (in urban as well as in rural areas) are likely to be the most vulnerable (Barnett and Webber 2010, p. 39). Scholars have identified two different aspects of the linkage between natural disasters and displacement (Collins 2013, p. 113): first, the intensity of the environmental shock or disaster is relevant when evaluating the possible impact on people's displacement; second, the vulnerabilities of the people of concern also need to be taken into consideration—including the given vulnerabilities pre-disaster, as well as post-disaster, in the short- and long-term.

Moreover, oftentimes people are unable to move, or they choose to stay put after being hit by an environmental shock. This phenomenon, which occurs during and after a natural disaster, is called 'entrapment' (Collins 2013, p. 113). An example of why some people choose not to leave—even in the course of a natural disaster—is that they hope to survive without giving up their possessions (Reuveny 2007, p. 658), and they often do not have enough financial resources at their disposal to migrate (IPCC 2012, p. 80). This is, among other case studies, illustrated by the 2004 Indian Ocean earthquake and tsunami. People affected by this disaster moved predominantly short distances and returned as soon as the government allowed it (Findlay 2011, p. 53). The floods in the 1990s around river deltas in Bangladesh showed a similar trend, as 88% of the people moved within a 2-mile zone (Collins 2013, p. 113). Also, case studies on Bangladesh have shown that many people do not move to cyclone shelters at all. After all, shelters often do not serve the basic needs, and people do not want to leave behind their possessions and farm animals. Additionally, Bangladeshis assigned religious meaning to the disaster and oftentimes did not take the necessary steps to escape the shock (Collins 2013, p. 113).

According to the 'IPCC Special Report', the number of people who were displaced because of rapid-onset climate-related disasters rose to around 20 million in 2008 alone (IPCC 2012, p. 80). The IPCC referred to findings from the UN Office for the Coordination of Humanitarian Affairs and the IDMC for this number. Here, the improved reporting of environmental extremes over time explains the increase in the number of people affected by disasters. Still, it is difficult to calculate the number of people displaced due to disasters because the International Disaster Database EM-DAT was only started in 2008 and has been influenced by a few large cases that have contributed 50% of the total number of people displaced (Black et al. 2013b,

p. 34).[10] With respect to displacement being a long-term phenomenon for the people affected, there is no dataset that depicts the number of people who move before or after the peak of a disaster, who return or who are still displaced after a certain number of years (Black et al. 2013b, p. 34). Overall, natural disasters tend to be worse in less developed countries and often are—in some form or another—worsened by the conditions of internal or external conflict (Enia 2008, p. 8).[11] In some cases, the fear of large-scale environmental displacement has triggered the involvement of external political actors in post-disaster reconstruction. Therefore, the question arises of why some natural disasters provide the space for conflict resolution, but others do not (Enia 2008, p. 9). The academic literature on displacement due to natural disasters has repeatedly mentioned certain events: In the years 2001, 2007 and 2008, floods in Mozambique displaced 200,000, 163,000 and 102,000 people, respectively, and changing flood patterns in the Mekong River Delta region also led to displacement (IPCC 2012, p. 80). Hurricane Katrina also is a common case study because the disaster led to the displacement of over one million people (Myers 2005).[12]

In a similar vein, another branch of the disaster literature has focused on the effect of poverty on disaster displacement. Empirically, we can see that economically impoverished people "often gather, or are gathered, in areas of higher environmental risk" (Collins 2013, p. 120), and they are at the mercy of environmental changes for this reason. As a consequence, scholars have supported poverty reduction, and they have strengthened the effects this might have on disaster vulnerability reduction (Alexander 1997, p. 287; Collins 2013, p. 120).[13] Also, the potential of labor migration as a strategy for adaption to environmental changes is discussed in this context (Barnett and Webber 2010, p. 42). Case studies on the topic have been extensive, usually focusing on a variety of drivers and impacts of environmental disasters. Bangladesh is one of the countries that has been given substantial academic attention, which is surely derived from the fact that it ranks among the world's most disaster-prone countries (Alam and Collins 2010, p. 932). Additionally, the

---

[10]As examples, the floods of May 2010 in China and of July and August 2010 in Pakistan can be mentioned. These contributed to over 10 million people becoming displaced (Black et al. 2013b, p. 34).

[11]For more information on 'disaster diplomacy', see Enia (2008). The author analyzed the transformative effect of the Tsunami of 2004 on internal conflicts in Indonesia and Sri Lanka with a special focus on post-disaster reconstruction by the government and/or opposition groups (Enia 2008, p. 16).

[12]Research on Hurricane Katrina has focused on social vulnerability, class and race, as well as on the questions of who received aid from government agencies and who was able to return after displacement (IPCC 2012, p. 80).

[13]The characteristics of vulnerability research, e.g. sensitivity, exposure and adaptive capacity, have emerged from the literature on risk-hazards and the food-security; meanwhile they have been included in the research on environmental changes (Luers 2005, p. 215). For a classification of the types of vulnerability, see Alexander (1997, p. 292), who differentiated between total vulnerability, economic vulnerability, technological vulnerability, newly generated vulnerability and residual unameliorated vulnerability.

countries affected by the 2004 Tsunami and Hurricane Katrina have served as case studies in much of the academic literature on disaster displacement (Enia 2008, p. 17). Further, because Pakistan suffered a series of earthquakes due to its geographic fault lines, the displacement in the country has also acted as an empirical case study (Enia 2008, p. 23).

The research has demonstrated a selectivity among those who return after a sudden-onset disaster—with the poorest people being least able to take this step (Findlay 2011, p. 53). The 2004 tsunami in the Indian Ocean illustrates this point quite clearly.[14] Studies have shown that 2 years after the tsunami, nearly half of those displaced could not return due to financial and political reasons and because of earlier conflicts in the region (Black et al. 2013b, p. 38). Therefore, a new body of literature has argued in favor of a human rights-based approach in environmental displacement studies. In essence, the case studies have shown that slow- as well as sudden-onset disasters are prone to human rights violations: Marginalized groups often suffer, first, under extreme vulnerability, and second, under disadvantages when it comes to post-disaster help (Kolmannskog 2012, p. 1075). The 'IPCC Special Report' specifically noted that "refugees, internally displaced people, and those in marginal areas as a result of violence can be dramatic examples of people vulnerable to the negative effects of weather and climate events, cut off from coping mechanisms and support networks" (IPCC 2012, p. 238). Authors in favor of the human rights-based approach have argued that this approach includes the people who left pre-disaster as a precaution and who therefore are often not attributed to environmental drivers (Kolmannskog 2012, p. 1077). Also, environmental changes directly impact human vulnerability, and a human rights-based approach better incorporates the interplay of socio-economic, political and environmental factors (Kolmannskog 2012, p. 1077). To gain a better insight into the phenomenon, current displacement research has also included gender aspects: According to the 'IPCC Special Report', displacement is "highly gendered in terms of both drivers and impacts" (IPCC 2012, p. 298), and this impact becomes apparent in the labor market and in the domestic sphere (Meertens and Segura-Escobar 1996, p. 173).

This thesis therefore follows Black et al. (2013b) and Piguet (2008), and differentiates between the diverse forms of displacement, e.g. temporary displacement of less than 3 months, short-term migration of 3 months to 1 year, and long-term migration of more than a year. Also, the factor of distance is important when comparing the displacement of people due to different kinds of natural disasters (Black et al. 2013b, p. 36). Acknowledging all these aspects, the work at hand follows Black et al., who understood displacement as "movement associated with discrete events that challenge safety, security or livelihoods. Much displacement is (...) involuntary or forced, and sometimes sudden if associated with rapid-onset hazards" (Black et al. 2011, p. 6). Movement is thus forced against the will of the individuals because they are compelled to move by events beyond their direct control (Collins 2013, p. 112; Couldrey and Herson 2014, p. 5). Here, displacement

[14]For a more detailed depiction of the displacement post-2004 tsunami, see Gray et al. (2014).

can be temporary as well as protracted over a longer period of time (Couldrey and Herson 2014, p. 5). The act of moving to merely improve livelihoods and opportunities is typically voluntary and has an implied act of planning in it (Black et al. 2011, p. 6). That said, the term human mobility is an umbrella term—with migration and displacement being forms of human mobility.

### 2.3.3 EDPs from the Perspective of Refugee Studies

The refugee crises of the 1990s led to the international spread of internal conflict, as exemplified in Central Africa, West Africa, the Middle East and the Balkans. Cases of large scale refugee flows are sometimes the starting point of ethnic conflicts—as in former Yugoslavia, the former Soviet Union, Indonesia and the DRC—and at times even trigger military intervention—as in Fiji and Cote d'Ivoire (Lischer 2006, p. 4). As of 2018, there are unresolved empirical questions on the refugee topic, such as: Under what conditions do refugee crises lead to the spread of civil war across borders? What role can and should government organizations play in preventing the spread of violence and conflict? And how do refugee flows affect the dynamics of conflict in various parts of the world? (cf. Lischer 2006, p. 2). In an attempt to give answers to these questions, research has focused on socio-economic explanations for refugee-related violence. Such studies have frequently mentioned the sizes of refugee camps, as well as the location of the camps (distance to the border of the home country). Additionally, poor living conditions combined with a large number of young men likely lead to militancy in refugee camps (Lischer 2006, p. 9). An important factor is whether states act responsibly in times of a refugee crisis or if they do not fulfill their obligation to provide security. Even though the studies on refugee-related violence have rarely focused on the type of violence that occurs in the camps, it is necessary to identify qualitative and quantitative differences among incidents (such as theft, rioting and international war) because the types of violence help to reconstruct the causes of violence in refugee camps.[15]

Nevertheless, refugee camps are just one example of how people displaced by disasters cope with their situation. The state of research on refugee camps can be summarized as mainly focusing on different socio-economic factors such as ethnicity, culture, reactions of host communities, coping mechanisms of the people in the camps, failures and successes by international aid agencies and small-scale violence (especially violence against women) in the camps (Horn 2010, p. 359). As an example, Reuveny argued that migration and displacement is generally conflict-prone due to factors such as competition (e.g., the arriving refugees might become a

[15]Lischer differentiated between the following types of violence in refugee camps: attacks between the sending state and the refugees, attacks between the receiving state and the refugees, ethnic or factional violence among the refugees, internal violence within the receiving state and inter-state war or unilateral intervention (Lischer 2006, pp. 11–12).

burden on the economy), ethnic tension (e.g., in cases where the residents of a refugee camp belong to different ethnic groups) and distrust (e.g., in cases of cross-border migration, the home country might suspect that the host country aims at upsetting the ethnic balance in the country of origin) (Reuveny 2007, p. 659).[16] Case studies so far have focused on Kenya because the country is host to several refugee camps, especially the Kakuma Refugee Camp (Aubone and Hernandez 2013, p. 32; Horn 2010, p. 359). Authors who specialize in this field of research have concluded that "it is impossible to quantify the amount of violence which takes place in and around Kenya's refugee camps" (Crisp 2000, p. 601). In addition to this, a very common effect of displacement and refugee life is "an increase in domestic violence" (Horn 2010, p. 356). This fact is repeatedly mentioned by refugees living in Kenya's Kakuma and Dadaab refugee camps.[17] Field research has shown that many refugees feel unsafe because their safety is under constant and serious threat. One Somali refugee in Kakuma recapped that "it is of no advantage for us to get a full ration from UNHCR if our lives are always at risk from insecurity" (Crisp 2000, p. 602). Additionally, the movement of people after war and conflict is unpredictable to some extent: Whereas some people move as soon as possible, others will try to stay in their homes (McAdam 2010, p. 2).

The above-mentioned aspects of refugee research therefore need to be acknowledged when dealing with EDPs. Also, certain elements of the 1951 Refugee Convention (see Sect. 5.1) apply to EDPs if they fall under the Convention's refugee definition. For example, the right of nonrefoulement implies that if refugees risk persecution in their home country, the host country is not allowed to forcibly send them back to their country. It is one of the cornerstones of global refugee protection. As all states have the obligation to grant refugees at least some kind of temporary protection on their territory, states neighboring refugee-producing countries often carry a comparatively greater responsibility for refugee protection (Loescher and Milner 2011, p. 192). In the refugee context, it is sometimes difficult to differentiate between voluntary migration and forced migration, and from the viewpoint of those affected, there is no reason to distinguish between climate change-related or natural disaster-related triggers (Kolmannskog 2012, p. 1072).

Even though scholars as well as practitioners have interpreted the occurrence of EDPs quite differently, a general trend in the debate has centered on aspects of human, national or international security (among others, Lischer 2006, p. 1; Methmann et al. 2013, p. 95). Myers' estimate of millions of 'climate refugees' by 2050 is one reason why the media and politicians of all shades have addressed the topic (Methmann et al. 2013, p. 95). In this context, security concerns include "cross-border incursions, militarization of refugee camps, and the fear of conflicts spilling

[16]For information on more factors, see Reuveny (2007, p. 659).

[17]Extensive research on violence in Kenyan refugee camps shows different forms of insecurity in refugee camps: domestic and community violence, sexual abuse and violence, armed robbery, violence within national refugee groups, violence between national refugee groups and violence between refugees and the local population (Crisp 2000, p. 604).

over from neighboring refugee-producing countries" (Lischer 2006, p. 1). The initial incentive of refugees to flee from violent conflict often leads to an expansion of violence because refugee crises are abused as strategies of war—e.g. as a pool of recruits, a source for food and medicine, a shield against attack and as international legitimacy for exiled groups (Lischer 2006, p. 2). Furthermore, the extent to which refugees become a burden on their host country (in cases of cross-border displacement) or host communities (in cases of internal displacement) depends heavily on the host's response (IPCC 2012, p. 399). The degree of political stability, the effectiveness of governance, the economy and the history of violence are relevant in this context (IPCC 2012, p. 399).

### *2.3.4 EDPs from the Perspective of Security Studies*

According to Scheffran et al., "climate change has no automatic effect on human security, on societal stability, or on violent conflict" (Scheffran et al. 2012a, p. 797). Scheffran et al. spotted multiple causal links between changes in the natural environment, on the one hand, and the phenomena that have effects on climate change, on the other hand. Especially important are the kinds of institutions—such as local town planning or forums for international negotiations—and their possible effects on different variables of the chain between environmental changes and conflict (Scheffran et al. 2012a, p. 797). A second conclusion of Scheffran et al. was that "climate change migration has often been seen only as a negative force—a threat to human security, a threat to societal stability, and a generator of armed conflict" (Scheffran et al. 2012a, p. 797). In contrast to this, migration also is a strategy of adaption to environmental changes. Similarly, Meierding argued that "scholars have failed to identify consistent relationships between climate change and conflict in systematic, cross-national analyses" (Meierding 2013, p. 185). Meierding's critique addressed Homer-Dixon's model, which conflict researchers have repeatedly used as their source of reference. In her opinion, the relationship between environmental changes and scarcity depends on as many intervening factors as the relationship between conflict and scarcity. As examples, Meierding mentioned social, economic and political factors as intervening variables, and she claimed that the existing variables lead to a multitude of links in the chain, thus generating an "already unwieldy model even less amenable to statistical testing" (Meierding 2013, p. 189).

But even though the current empirical research seems to have questioned the security implications of environmental changes, some academics have argued that the public still perceives the environment as a security issue. Here, discursive analyses have illustrated that the storyline of a threat of environmental migration has been successful to some extent (cf. among others, Piguet 2013, p. 154). Nevertheless, the security references have not resulted in exceptional measures at the political level as assumed by the Copenhagen School (Rothe 2012, p. 243). Researchers such as Brauch (2012), Brzoska (2009), and Trombetta (2014) have applied the Copenhagen School approach thoroughly, and they have all come to the conclusion that "things in the empirical field of climate change politics are more complex than suggested in the framework" (Rothe 2012, p. 243). Rothe (2012) highlighted the fact that there is not *one* security discourse in the field of environmental changes and security but multiple concepts. The development of a "macro-securitization" (Rothe 2012, p. 254) depicts climate change as an overall threat to humankind and thereby intensifies the politization of environmental politics. As a possible side-effect, Rothe saw a "de-politization of international climate governance" (Rothe 2012, p. 254).

With respect to large-n empirical studies, Butler and Gates declared that few if any of these studies had established a link between scarcity and interstate or intrastate war (Butler and Gates 2012). They focused on pastoral range wars in Africa and elucidated some cases that might support the scarcity argument for intercommunal violence. Nevertheless, empirically based case studies have contradicted these cases (Butler and Gates 2012, p. 25). By the same token, system level analyses have used the counts of militarized disputes (Militarized Interstate Dispute dataset) and have combined them with other variables such as the annual average temperature, GDP numbers and regime type (Policy IV project) (Gartzke 2012, pp. 183–184). These new research approaches have aimed at producing data sets on a smaller scale and on non-state conflicts that are associated with environmental change (Meierding 2013, p. 186). Another recent development is the use of geographically disaggregated data to identify correlations between environmental change and conflict. This implies that researchers have geographically located conflicts and have collected environmental, social and economic data at the subnational level (Meierding 2013, p. 186). Yet another development in the current research has been the use of two-stage statistical analysis techniques and an intermediate variable—e.g. economic growth—to model the indirect effects between environmental change and conflict (Meierding 2013, p. 186). The slums of Dhaka in Bangladesh have served as one example of how the linkage between environmental change and social stresses on the ground might pose a challenge to peace and security (Saha 2012, p. 606).

Nevertheless, scholars such as Rothe (2012) have emphasized the need for studies on the political impacts of a securitization of environmental changes that cover a longer time span. The aim would be to uncover the long-term effects of the securitization process and to "include a wider range of policy fields such as development or security politics to account for possible spill-over effects and interactions with international climate politics" (Rothe 2012, p. 254). Similarly, Kjellen and Wallensteen introduced a new concept called the "new diplomacy for sustainable development" (Kjellen and Wallensteen 2012, p. 686), arguing that policymakers and researchers should accept that one cannot negotiate or use traditional bargaining techniques with environmental changes. It requires a new perception of the nature of the threat that incorporates a long-term perspective and includes accepting the problem as a societal issue. Concerning international negotiations and conventions, the different countries should focus on the bargaining position of other parties to achieve better results in the negotiations on climate change (Kjellen and Wallensteen 2012, p. 686).

Adding to this and concerning the linkage between environmental changes and conflict, there has been a critique of the re-use of preexisting quantitative data. Instead, new data sets such as the Social Conflict in Africa Database, the UCDP Non-State Conflict Dataset, and the country-specific conflict data collection efforts could enhance academic studies (Meierding 2013, p. 190). Further, some have argued that the economic growth variable should be substituted for a focus on agriculture: According to Meierding, the "agricultural sector is most likely to be influenced by climate change" (Meierding 2013, p. 186). Therefore, local institutions could be used as examples of how additional variables can interrupt the causal chain between environmental changes and conflict. In addition, Butler and Gates argued that it is problematic that the focus of most research has been on intra- or interstate conflict. Rather, the focus should be on intercommunal or local ethnic conflicts—where the link between environmental scarcity and conflict is most apparent (Butler and Gates 2012, p. 24). The institutions at the community level are suitable as research objects because they "are likely to be the mechanisms that individuals and communities employ to manage local environmental degradation and maldistribution" (Meierding 2013, p. 195). Examples of these non-state-level institutions are local courts, conflict resolution devices, treaties, social norms and customary practice. The literature on the link between environmental changes and conflict has only recently shown development in these local institutional mechanisms.

Additionally, researchers have called for more efforts to approach environmental changes; if they lead to the economic stagnation of the countries affected by climate change, this might indeed result in climate change conflicts (Gartzke 2012, p. 178). In a similar vein, the research needs to address both the hypothesis that rising temperatures increase conflict on a global scale as well as rule out the possibility that the causes of climate change in themselves increase violent conflicts (Gartzke 2012, pp. 178–179). The findings from disaster sociology research indicate that disasters have no significant impact on violent conflicts and that people pull together in the wake of a natural disaster instead of fighting each other (Meierding 2013, p. 196). Also, the main characteristic of a disaster is its uncertainty concerning the occurrence and degree of damage. The incorporation of this kind of uncertainty into the research on climate change and conflict will make a valuable contribution to conflict research (Meierding 2013, p. 200).

To conclude, this chapter has established the case and it has presented the propositions that security studies make concerning EDPs. This chapter's focus on security references has thereby outlined the linkage between security and environmental changes. Additionally, this chapter introduced environmental changes and the prominent interpretations of environmental security and environmental conflict. In a last step, this chapter has looked at the topic of EDPs from the perspectives of migration studies, disaster displacement studies, refugee studies and security studies. Hereby, the predominant scholarly debates on EDPs became apparent. This chapter is the basis for the theoretical approach which will be outlined in the next chapter.

## References


Alam E, Collins AE (2010) Cyclone disaster vulnerability and response experiences in coastal Bangladesh. Disasters 34:931–954

Alexander D (1997) The study of Natural Disasters, 1977-1997: some reflections on a changing field of knowledge. Disasters 21:284–304

Aubone A, Hernandez J (2013) Assessing refugee camp characteristics and the occurrence of sexual violence: a preliminary analysis of the Dadaab Complex. Refug Surv Q 32:22–40. https://doi.org/10.1093/rsq/hdt015

Baldwin A, Bettini G (2017) Introduction: Life adrift. In: Baldwin A, Bettini G (eds) Life adrift: climate change, migration, critique. Rowman & Littlefield, Washington, DC, pp 1–22

Balzacq T (2005) The three faces of securitization: political agency, audience and context. Eur J Int Relat 11:171–201

Barnett J, Webber M (2010) Migration as adaptation: opportunities and limits. In: McAdam J (ed) Climate change and displacement. Hart, Oxford, pp 37–56

Bigo D (2000) When two become one: internal and external securitisations in Europe. In: Kelstrup M, Williams MC (eds) International relations theory and the politics of European integration: power, security and community. Routledge, New York, pp 171–204

Bigo D (2002) Security and immigration: toward a critique of the governmentality of unease. Alternatives 27:63–92

Black R, Adger WN, Arnell NW, Dercon S, Geddes A, Thomas D (2011) The effect of environmental change on human migration. Glob Environ Change 21:3–11. https://doi.org/10.1016/j.gloenvcha.2011.10.001

Black R, Adger WN, Arnell NW (2013a) Migration and extreme environmental events: new agendas for global change research. Environ Sci Policy 27:1–3. https://doi.org/10.1016/j.envsci.2012.09.010

Black R, Arnell NW, Adger WN, Thomas D, Geddes A (2013b) Migration, immobility and displacement outcomes following extreme events. Environ Sci Policy 27:S32–S43. https://doi.org/10.1016/j.envsci.2012.09.001

Boas I, Rothe D (2016) From conflict to resilience? Explaining recent changes in climate security discourse and practice. Environ Polit 25:613–632

Booth K (1991) Security and emancipation. Rev Int Stud 17:313–326

Bourbeau P (2014) Moving forward together: logics of the securitisation process. Millennium J Int Stud 43:187–206. https://doi.org/10.1177/0305829814541504

Brauch HG (2012) Policy responses to climate change in the Mediterranean and MENA region during the Anthropocene. In: Scheffran J, Brzoska M, Brauch HG, Link PM, Schilling J (eds) Climate change, human security and violent conflict: challenges for societal stability, vol 8. Springer, New York, pp 719–794

Brzoska M (2009) The securitization of climate change and the power of conceptions of security. Secur Peace 27:137–145. https://doi.org/10.5771/0175-274x-2009-3-137

Butler CK, Gates S (2012) African range wars: climate, conflict, and property rights. J Peace Res 49:23–34. https://doi.org/10.1177/0022343311426166

Buzan B, Wæver O, Jd W (1998) Security: a new framework for analysis. Rienner, Boulder, CO

Collins AE (2013) Applications of the disaster risk reduction approach to migration influenced by environmental change. Environ Sci Policy 27:112–125. https://doi.org/10.1016/j.envsci.2012.10.005

Corry O (2012) Securitisation and 'riskification': second-order security and the politics of climate change. Millennium J Int Stud 40:235–258. https://doi.org/10.1177/0305829811419444

Couldrey M, Herson M (2014) Crisis. Forced Migr Rev 45:1–100

Crisp J (2000) A state of insecurity: the political economy of violence in Kenya's refugee camps. Afr Aff 99:601–632

Elliott L (2010) Climate migration and climate migrants: what threat, whose security? In: McAdam J (ed) Climate change and displacement. Hart, Oxford, pp 175–190

Enia J (2008) Peace in its wake? The 2004 Tsunami and internal conflict in Indonesia and Sri Lanka. J Public Int Affairs 19:7–27
Findlay AM (2011) Migrant destinations in an era of environmental change. Glob Environ Change 21:50–58. https://doi.org/10.1016/j.gloenvcha.2011.09.004
Gartzke E (2012) Could climate change precipitate peace? J Peace Res 49:177–192. https://doi.org/10.1177/0022343311427342
Geddes A, Somerville W (2012) Migration and environmental change in international governance: the case of the European Union. Environ Plan Gov Policy 30:1015–1028
Gillespie T, Chu J, Frankenberg E, Thomas D (2007) Assessment and prediction of natural hazards from satellite imagery. Prog Phys Geogr 31:459–470
Gray C, Frankenberg E, Gillespie T, Sumantri C, Thomas D (2014) Studying displacement after a disaster using large-scale survey methods: Sumatra after the 2004 Tsunami. Ann Assoc Am Geogr 104:594–612
Helm D, Hepburn C (2009) The economics and politics of climate change. Oxford University Press, Oxford
Homer-Dixon TF (1991) On the threshold: environmental changes as causes of acute conflict. Int Secur 16:76–116
Homer-Dixon TF (1994) Environmental scarcities and violent conflict: evidence from cases. Int Secur 19:5–40
Horn R (2010) Exploring the impact of displacement and encampment on domestic violence in Kakuma Refugee Camp. J Refug Stud 23:356–376. https://doi.org/10.1093/jrs/feq020
Huysmans J (2011) What's in an act? On security speech acts and little security nothings. Security Dialogue 42:371–383
IPCC (2001) IPCC Third Assessment Report – Working Group 1: The scientific basis: Appendix 1 – Glossary ("climate change"). https://www.ipcc.ch/ipccreports/tar/wg1/pdf/WGI_TAR_full_report.pdf. Accessed 13 Mar 2019
IPCC (2012) Managing the risks of extreme events and disasters to advance climate change adaptation: a special report of Working Group I and II of the Intergovernmental Panel on Climate Change. https://www.ipcc.ch/pdf/special-reports/srex/SREX_Full_Report.pdf. Accessed 13 Mar 2019
Jackson NJ (2006) International organizations, security dichotomies and the trafficking of persons and narcotics in post-Soviet Central Asia: a critique of the securitization framework. Secur Dialogue 37:299–317
Kaldor M (2007) Human security. Polity, Cambridge
Kita SM, Raleigh C (2018) Environmental migration and international political security: rhetoric, reality and questions. In: McLeman R, Gemenne F (eds) Routledge handbook of environmental displacement and migration. Routledge, New York, pp 356–370
Kjellen B, Wallensteen P (2012) Climate change, peacekeeping, and perspectives for UN reform. In: Scheffran J, Brzoska M, Brauch HG, Link PM, Schilling J (eds) Climate change, human security and violent conflict: challenges for societal stability. Springer, New York, pp 685–694
Kniveton D, Smith C, Wood S (2011) Agent-based model simulations of future changes in migration flows for Burkina Faso. Glob Environ Change 21:34–40. https://doi.org/10.1016/j.gloenvcha.2011.09.006
Kolmannskog V (2012) Climate change, environmental displacement and international law. J Int Dev 24:1071–1081. https://doi.org/10.1002/jid.2888
Kurtz G (2012) Securitization of climate change in the United Nations 2007-2010. In: Scheffran J, Brzoska M, Brauch HG, Link PM, Schilling J (eds) Climate change, human security and violent conflict: challenges for societal stability. Springer, New York, pp 669–684
Lennon AT, Smith J (2007) The climate for terror. The New York Times, New York. https://www.nytimes.com/2007/12/03/opinion/03iht-edsmith.1.8569518.html. Accessed 1 July 2019
Lenton TM (2013) What early warning systems are there for environmental shocks? Environ Sci Policy 27:60–75. https://doi.org/10.1016/j.envsci.2012.06.011

Léonard S, Kaunert C (2011) Reconceptualizing the audience in securitization theory. In: Balzacq T (ed) Securitization theory: how security problems emerge and dissolve. Routledge, New York, pp 57–76

Lischer SK (2006) Dangerous sanctuaries: refugee camps, civil war, and the dilemmas of humanitarian aid. Cornell University Press; University Presses Marketing, Ithaca, NY, Bristol

Loescher G, Milner J (2011) UNHCR and the global governance of refugees. In: Betts A (ed) Global migration governance. Oxford University Press, New York, pp 189–209

Luers AL (2005) The surface of vulnerability: an analytical framework for examining environmental change. Global Environ Change 15:214–223. https://doi.org/10.1016/j.gloenvcha.2005.04.003

McAdam J (2010) Introduction. In: McAdam J (ed) Climate change and displacement. Hart, Oxford, pp 1–8

Meertens D, Segura-Escobar N (1996) Uprooted lives: gender, violence and displacement in Colombia. Singap J Trop Geogr 17:165–178

Meierding E (2013) Climate change and conflict: avoiding small talk about the weather. Int Stud Rev 15:185–203. https://doi.org/10.1111/misr.12030

Methmann C, Rothe D, Stephan B (eds) (2013) Interpretive approaches to global climate governance: (de)constructing the greenhouse. Routledge, New York

Myers N (2005) Environmental refugees: an emergent security issue. Session III – Environment and migration. https://www.osce.org/eea/14851?download=true. Accessed 13 Mar 2019

Nordas R, Gleditsch NP (2007) Climate change and conflict. Polit Geogr 26:627–638

Oels A (2012) From "securitization" of climate change to "climatization" of the security field: comparing three theoretical perspectives. In: Scheffran J, Brzoska M, Brauch HG, Link PM, Schilling J (eds) Climate change, human security and violent conflict: challenges for societal stability. Springer, New York, pp 185–205

Oels A, von Lucke F (2015) Gescheiterte Versicherheitlichung oder Sicherheit im Wandel: Hilft uns die Kopenhagener Schule beim Thema Klimawandel? Zeitschrift für Internationale Beziehungen 22:43–70

Piguet E (2008) Climate change and forced migration. New Issues in Refugee Research, Geneva

Piguet E (2013) From "primitive migration" to "climate refugees": the curious fate of the natural environment in migration studies. Ann Assoc Am Geogr 103:148–162

Reusswig F, Lass W (2013) Klimaschutz als globale Norm und soziale Wirklichkeit. In: Pinéu F d R, Andrea KM, Schapper A (eds) Globale Normen zwischen Anspruch und Wirklichkeit. Nomos, Baden-Baden, pp 231–256

Reuveny R (2007) Climate change-induced migration and violent conflict. Polit Geogr 26:656–673. https://doi.org/10.1016/j.polgeo.2007.05.001

Rosenow-Williams K, Behmer K (2015) Gendered environmental security in IDP and refugee camps. Peace Rev J Soc Justice 27:188–195

Rothe D (2012) Security as a weapon: how cataclysm discourses frame international climate negotiations. In: Scheffran J, Brzoska M, Brauch HG, Link PM, Schilling J (eds) Climate change, human security and violent conflict: challenges for societal stability, vol 8. Springer, New York, pp 243–258

Rothe D (2013) A climate of insecurity: the securitisation of global warming and its policy implications. Dissertation am Fachbereich Sozialwissenschaften der Universität Hamburg, Hamburg

Rothe D (2016) Securitizing global warming: a climate of complexity. Routledge, New York

Saha S (2012) Security implications of climate refugees in urban slums: a case study from Dhaka, Bangladesh. In: Scheffran J, Brzoska M, Brauch HG, Link PM, Schilling J (eds) Climate change, human security and violent conflict: challenges for societal stability, vol 8. Springer, New York, pp 595–611

Scheffran J, Brzoska M, Brauch HG, Link PM, Schilling J (eds) (2012a) Climate change, human security and violent conflict: challenges for societal stability, vol 8. Springer, New York

Scheffran J, Brzoska M, Kominek J, Link PM, Schilling J (2012b) Disentangling the climate-conflict nexus: empirical and theoretical assessment of vulnerabilities and pathways. Rev Eur Stud 4:1–13

Schilling J, Freier KP, Hertig E, Scheffran J (2012) Climate change, vulnerability and adaption in North Africa with focus on Morocco. Agric Ecosyst Environ 156:12–26

Schlesinger M (2008) Human security vs. collective security: an empirical study of security council resolutions. VDM Verlag Dr. Müller, Saarbrücken

Tadjbakhsh S, Chenoy A (2007) Human security: concepts and implications. Routledge advances in international relations and global politics. Routledge, New York

Tänzler D, Ries F (2012) International climate change policies: the potential relevance of REDD+ for peace and stability. In: Scheffran J, Brzoska M, Brauch HG, Link PM, Schilling J (eds) Climate change, human security and violent conflict: challenges for societal stability, vol 8. Springer, New York, pp 695–705

Trombetta MJ (2008) Environmental security and climate change: analysing the discourse. Camb Rev Int Aff 21:585–602. https://doi.org/10.1080/09557570802452920

Trombetta MJ (2014) Linking climate-induced migration and security within the EU: insights from the securitization debate. Crit Stud Secur 2:131–147. https://doi.org/10.1080/21624887.2014.923699

UN Secretary-General (2004) A more secure world: our shared responsibility. Report of the high-level panel on threats, challenges and change. New York, United Nations

UN General Assembly (2009) Climate change and its possible security implications. Report of the Secretary-General. New York, UN General Assembly. https://www.unhcr.org/protection/environment/543e73f69/climate-change-its-possible-security-implications-report-secretary-general.html. Accessed 1 July 2019

von Lucke F, Wellmann Z, Diez T (2014) What's at stake in securitising climate change? Towards a differentiated approach. Geopolitics 19:857–884. https://doi.org/10.1080/14650045.2014.913028

Warner K (2010) Global environmental change and migration: governance challenges. Glob Environ Change 20:402–413. https://doi.org/10.1016/j.gloenvcha.2009.12.001

Warner K, Hamza M, Oliver-Smith A, Renaud F, Julca A (2010) Climate change, environmental degradation and migration. Nat Hazards 55:689–715. https://doi.org/10.1007/s11069-009-9419-7

Webersik C (2010) Climate change and security: a gathering storm of global challenges. Security and the environment. Praeger, Santa Barbara, CA

# Chapter 3
# Theoretical Background: Organizational Theories and the Research on International Organizations

This chapter builds upon the previous chapter that has established the case of this thesis. With regard to the background of EDPs as discussed in Chap. 2, this chapter takes a theoretical stance and discusses approaches from organizational theory to address the above-mentioned research gap. In order to answer the research question of why the UNHCR approaches the topic of EDPs the way that it does, the analysis will be guided by a classification of organizational theory into internal and external factors.

To begin with, it is necessary to define what is meant by the term 'international organization' in the context of this thesis. The term 'organization' is often used in three different contexts: First, it is used in the context of unions and associations. Second, it refers to processes that aim to regulate actions in order to achieve a certain purpose or goal. In other words, activities that were previously unrelated are put into a meaningful order, and thus, represent the rational outcomes of preceding decision-making (Weick 1985, p. 11). As such, the act of organizing can be compared to grammar because both imply the systematic combination of rules and conventions in order to make certain social processes more comprehensible for the people involved (Weick 1985, p. 12). And, third, an organization can be the end product of these organizing acts and refer to a certain order or regime (cf. Mayntz 1963; Rittberger and Zangl 2003, p. 22). As such, the acts of IOs evolved in modern society at the same time as—among other things—bureaucratic administration, standing armies and professional soldiers, and educational systems such as schools and universities (Kühl 2011, pp. 15–16). Building upon Chap. 1, this thesis defines an IO as a purposeful and (mostly) autonomous and legitimacy-driven actor that moves in the international arena and that has a definable membership and a hierarchical structure. The thesis hereby uses the term 'research on IOs' for the research on international organizations and the term 'organizational research' for the research on organizational theories.



S. Hantscher, *The UNHCR and Disaster Displacement in the 21st Century*,
Contributions to Political Science, https://doi.org/10.1007/978-3-030-19689-9_3

## 3.1 Critical Discussion of International Relations Theory

The current state of research says that (Neo-)Realism, as well as (Neo-)Liberalism, cannot elucidate important organizational behavior (cf. Abbott and Snidal 1998, p. 8; Barnett and Finnemore 1999, pp. 699–702; Dijkzeul and Beigbeder 2003, p. 15; Nielson and Tierney 2003, pp. 243–244; Steffek 2015, p. 265; Weaver and Leiteritz 2005, p. 369). Even though the theoretical background of Realism helps to explain why states created the UNHCR and how the Realpolitik and power politics might reflect in the organization's policies, Realism is nevertheless insufficient for grasping why states give responsibility and power to the UNHCR. From the perspective of Realism, IOs simply represent power constellations in the international system, and IOs do not have authority over their member states. Realism is inapplicable because it underestimates the role of and attaches no importance to IOs, and in its state-centric understanding, IOs are only kept alive if states benefit from them or exercise power through them. In other words, "states use international institutions to further their own goals, and they design institutions accordingly" (Koremenos et al. 2001, p. 762). From the perspective of Realism, "IOs are simply handmaidens of powerful states and hence cannot formulate a global public interest" (Steffek 2015, p. 265). Moreover, Realists' prediction of a falling apart of IOs did not prove true; instead, IOs pursued taking center stage in global governance and even rose in importance after the Cold War (Kleine 2013b, p. 325). Rationalists use an economic argument to stress that IOs resolve the transaction costs for states and thus help to solve collective action problems (Neumann and Sending 2013, pp. 134–135). That said, Rationalists see IOs as having a certain autonomy, although this depends upon influential states inside an IO (Neumann and Sending 2013, pp. 134–135). In the end, Rationalists "believe that the answer lies in states" (Johnson 2013, p. 183). The problem with applying any kind of realist or rationalist theory to an IO is that "these theories provide no ontological independence to IOs, they have no way to test for autonomy nor have they any theoretical cause or inclination to test for it since, by theoretical axiom, autonomy cannot exist" (Barnett and Finnemore 1999, pp. 714–715). Weaver and Leiteritz consent to this critique: They declare realist and rational institutionalist approaches insufficient to explain organizational change, because these theories simply assume that states adapt IOs as they wish (Weaver and Leiteritz 2005, p. 370). The broad range of liberal theories of IR shed light on some of the black spots left behind by Realism (Kerler 2007, pp. 134–135). Nevertheless, the liberal interpretations of the workings of IOs are unsatisfactory due to their limited ability to open the 'black box' of the UNHCR. In other words, Liberalism is not able to illuminate the inner-organizational workings of IOs, and it thus does not give sufficient answers to the question of why the UNHCR approaches an issue that lies outside of its mandate. Therefore, Liberalism is unsuitable because its ontology entails that IOs only matter "as structural constraints on state behavior, not as autonomous actors" (Nielson and Tierney 2003, pp. 243–244). From the perspective of regime theory, IOs are the combination of norms and rules which all actors involved could agree upon. As such, IOs function to provide for common interest outcomes and to stabilize behavior between states (Kerler 2007, pp. 134–135). However, the newly developed analytical concept of international regimes was only a small step toward the recognition of IOs.

In contrary to the above-mentioned theoretical 'lenses' of IR, Constructivism is well suited to interpreting why the UNHCR uses security references as regards EDPs. Also, this theoretical 'lens' explains why the term 'climate refugees' is symbolically charged, and it asserts that this is relevant because—according to constructivism—ideas shape the social reality and thus change the way the world looks at the topic of EDPs. Constructivists point out that IOs can very well be called autonomous as opposed to their member states. Concerning their in-depth knowledge and bureaucratic features, IOs might sometimes even play an authoritative role toward states (Neumann and Sending 2013, pp. 134–135). The constructivist perspective acknowledges organizational change, but only to the extent to which it originates in and reflects "ideational shifts in epistemic communities and discursive structures constituting the normative environment of IOs" (Weaver and Leiteritz 2005, p. 370). More specifically, Social Constructivists argue that IOs look the way they do due to the international bureaucrats who transform the respective organizations from within (Johnson 2013, p. 183).[1] Nielson et al. (2006) point at the still-predominant divide between the rationalist and constructivist approaches to IOs. Instead of following research designs that analyze IOs and organizational change, either from a rationalist or a constructivist point of view, Nielson et al. propagate a synthesis of the two theoretical perspectives when working on an empirical case study (Nielson et al. 2006, pp. 108–109). The intention is to answer the questions regarding an IO's policy preferences that the rationalist approach is unable to tackle: "Unless an international organization's very survival is under threat, it is often difficult to attach a clear-cut material interest and an ensuing policy preference to an international organization" (Reinalda and Verbeek 2004, p. 234). Consequently, the constructivist approaches gain momentum and are necessary to analyze ideas and norms inside an IO, e.g. the specific positions of the IO's staff (Reinalda and Verbeek 2004, p. 234). In a similar vein, van der Heijden emphasizes a pluralistic approach to organizational research, especially when it comes to questions of organizational reform (van der Heijden 2012, p. 57).

This thesis concurs with recent research that questions the applicability of IR theory in the research on IOs (Knill et al. 2017b, p. 174; Steffek 2015, pp. 277–280). In contrast to the realist perspective that IOs "do not have any independent political agency" (Steffek 2015, p. 278), the following sections will present the perspectives of the research on IOs and organizational research. By this means, the sections introduce a different theoretical approach to IOs as compared to IR theory.

[1]This thesis concurs with the criticism of the existing research concerning how IOs should be tackled from a political science perspective: Ege and Bauer (2013) criticize the well-known and established approaches by scholars such as Barnett and Finnemore (2004) and argue that "bureaucratic characteristics (be they formal or informal) and the interaction with the worlds of politics remain conceptually blurred" (Ege and Bauer 2013, pp. 142–143).

## 3.2 The Research on International Organizations

### 3.2.1 *Defining International Organizations*

In research on IOs, the definitions of IOs often build on the three characteristics introduced by Niklas Luhmann—membership, purpose, and hierarchy (Luhmann 1976, pp. 29–39). The first element, membership, can be either universal or particular (Rittberger and Zangl 2003, p. 29) and implies that an IO has the authority over its members to decide (Luhmann 1976, pp. 39–53). An organization constitutes a space in which its members must follow the rules and meet the obligations established by the IO. In case an IO's constituent fails to live up to its membership conditions, it is possible for the membership to be recalled (Luhmann 1976, pp. 39–53). To specify even further, an IO is made up of sovereign states (more precisely, of more than three states), in contrast to private IOs, which allow a broader variety of parties (Bauer and Knill 2007; Dingwerth et al. 2009). The second characteristic of Luhmann is an organization's goal or purpose. Luhmann explains that the target-oriented character of some organizations is similar to the character of a company: Organizations provide commodities or services in order to make a profit. Other IOs have a rather all-encompassing purpose, which means that they deal with a variety of problem areas (Rittberger and Zangl 2003, p. 22). The absence of a purpose irritates an organization's constituents as well as the organizational environment (Luhmann 1976, pp. 100–104). Additionally, to be considered an IO, the member states must have installed a formal agreement and a permanent secretariat (Bauer and Knill 2007, pp. 13–14). The third element of Luhmann's definition is hierarchy (Luhmann 1976, pp. 156–162). Although modern societies are not exclusively organized in hierarchical terms, organizations are very hierarchical structures, irrespective of whether the IO acts for economic, political or academic purposes (Kühl 2011, p. 20). When capturing IOs in terms of membership, purpose and hierarchy, another important characteristic becomes apparent—autonomy. Following Luhmann, IOs decide autonomously about purpose, hierarchy and membership. Therefore, if an IO's autonomy is restricted in at least one of the three characteristics, it no longer constitutes an IO (Kühl 2011, p. 21). Bauer and Knill stress that organizational autonomy is necessary to run intra-organizational affairs (Bauer and Knill 2007). Actorness thus becomes important when engaging with bureaucracies (Mehde 2007, p. 163).

Furthermore, questions of legitimacy are relevant in research on IOs. Scholars assume that questions of IO legitimacy frequently come up in highly controversial or political circumstances. The increasing interest in the legitimacy of IOs simultaneously represents an increase in the role of IOs in global governance overall (Mehde 2007, p. 164). In the context of legitimacy, the democratic deficit of IOs is addressed. This implies that IOs lack a democratic character, e.g. an identifiable polity in which all members are able to participate in the public discourse (Zürn 1998, pp. 25–26). To make matters worse, the decision-making processes inside IOs are largely non-transparent, e.g. the decisions of IOs can rarely be traced back to the responsible office-bearer (Zürn 1998, pp. 25–26). Mehde's interpretation of the internal legitimacy of IOs refers to whether an IO takes the wishes of its constituents into consideration and acts accordingly; he equates external legitimacy with the impact of an IO's concrete actions on its member states or the wider public (Mehde 2007, pp. 164–165). Adding to this, some scholars have also pointed out that the legitimacy of IOs has to be viewed in a triangulation with costs and performance. Member states will only pay the costs of a potential loss of sovereignty and a membership fee if an IO performs according to its member states' wishes. This triangular process ultimately results in the increased legitimacy of the organization (Mehde 2007, p. 166). According to this understanding of the legitimacy of IOs, an organization is thus highly dependent on whether its member states perceive the IO's work as satisfactory. Vice versa, one may conclude that IOs are always challenged by having to live up to the expectations of multiple constituents (Mehde 2007, p. 173). Steffek (2015) contributes to the discussion about the legitimacy of IOs by furthering the understanding of output legitimacy of IOs. He argues that "IOs can avail of democratic output legitimacy to the extent that they keep powerful states in check and secure a high epistemic quality of their decision-making" (Steffek 2015, p. 265).

For the aim of this thesis, an IO is thus understood as a purposeful and (mostly) autonomous and legitimacy-driven actor that moves in the international arena and has a definable membership and a hierarchical structure. That said, the work at hand accords with the understanding of IOs as entities, compared to international institutions, which are regarded as (sets of) rules (Martin and Simmons 2013, p. 326).

### 3.2.2 *Capturing International Organizations*

Before introducing the various strands of organizational research, the common perceptions of IOs as arenas, agents or bureaucracies are presented (cf. Dingwerth et al. 2009, pp. 16–17; Hanrieder 2014b, p. 221; Neumann and Sending 2013, pp. 132–133) Rittberger and Zangl 2003, pp. 23–24). First, IOs are described by the use of the metaphor of an arena in which states operate and organize their intergovernmental politics. According to this interpretation, an IO is thus merely seen as presenting an institutional frame for interstate relations (Koremenos et al. 2001, p. 762). From a critical perspective, IOs are thus lacking any kind of autonomous role in global governance.

Second, IOs are seen as agents and are thus given a more prominent role in decision-making. According to the agent metaphor, states delegate orders to IOs, which thus enhances the status of the agents compared to the principals (states). However, this principal-agent approach still centers on states' involvement with IOs, and the organizations themselves take a backseat (Dingwerth et al. 2009, pp. 16–17). Here, the presumable rise in importance of IOs is ascribed to the states' instrumentalization of IOs, instead of a de facto rise in the relevance of IOs (Dingwerth et al. 2009, p. 18).

Third, as introduced by Barnett and Finnemore (2004) and in response to the perception of IOs as arenas or as agents, IOs are credited with being bureaucracies. This theoretical perspective provides an insight into IO pathologies (for a cutting-edge account on pathologies, see Barnett and Finnemore 2004 and into organizational culture e.g. Enkler and Grohs 2017 and Weaver and Leiteritz 2005, p. 369). Additionally, the approach of Haas falls under this category of organizational perception because he introduces a theoretical model that juxtaposes the adaptation and learning processes of bureaucracies (Haas 1990, pp. 6–7). In a similar vein, Siebenhüner differentiates between reflexive learning, adaptive learning and no learning, and analyzes the influences on learning in IOs (Siebenhüner 2008, pp. 96–98).

Other metaphorical connotations refer to IOs either as machines, clockworks or robots to stress an organization's predictability (Kühl 2011, p. 89). The metaphor of the machine is probably the oldest and means that there is something like an ideal type of organization, or that at least there is an optimal machine for each and every clientele. An IO understood as a game accentuates that there is a lot going on besides the organizational rules and standards. This metaphor thus highlights deviating behavior inside IOs and how difficult it is to predict an IO's decision-making. Additionally, IOs are sometimes described as theaters or theater stages, which implies that they try to gain support by means of a straightened external presentation (Kühl 2011, p. 89). By the same token, the idea of an IO as a façade focuses on the way the organization tries to present itself vis-à-vis its constituents in order to gain legitimacy (Kühl 2011, pp. 90–91).

## 3.3 A Classification of Organizational Theories

This thesis' approach builds on the various organizational theories in a complementary manner. As captured by the different metaphors mentioned above, the understanding of IOs usually focuses on either internal or external aspects, and it thus either builds on public administration research (internal) or on theoretical concepts of IR (external). That said, in order to be able to answer the underlying research question, it is paramount to take into consideration internal as well as external factors that might—in varying strengths and occurrences—influence the UNHCR's actions. While analyzing an IO's behaviors, certain theoretical strands will probably be more

**Table 3.1** Depiction of the current state of organizational research and classification of internal and external factors

| | Research that accentuates the organizational environment | Research that accentuates the role of states in organizations | Research that accentuates the character of the bureaucracy | Research that accentuates organizational strategies |
|---|---|---|---|---|
| Internal | | | X | X |
| External | X | X | | |

helpful than others, but the aim is to remain open to complementary factors that may have led to the observable result.

To improve readability, the following sections are organized in a 'top-down' manner, that is, from the organizational theory that focuses on external factors to the organizational theory that focuses on the internal elements of an IO (see Table 3.1). The aim is to subsume the state of the art on organizational research in a structured theory-led way: first, from theoretical approaches that explain organizational behavior with reference to global governance, the international environment (e.g. inter-organizational relations), and external factors; second, to theoretical concepts that emphasize the role of states in the actions of IOs; and third, to theoretical approaches that bring into focus the internal organizational structures, organizational culture and DNA, and the role of the leadership in IOs.

### *3.3.1 The 'Organizational Environment' Approach*

Representatives of this research strand advocate a complex understanding of organizations that exceeds the principal-agent and the bureaucracy approaches. That said, the aim of research that falls under this category is to differentiate between the various structures around organizations. This allows for research that considers the environment of IOs, such as the respective national or international context. To start with, this strand of research acknowledges that IOs must do the splits: On the one hand, organizations are confronted with an ever increasing and more heterogeneous circle of addressees (due to the rise in member states); on the other hand, they are faced with a social environment that contributes to stronger politicization (e.g. due to the civil society and sponsors) (Dingwerth et al. 2014, p. 186).

This category of organizational approaches draws on scholars such as Dingwerth et al. (2009, p. 15) and Meyer and Rowan (1991, pp. 49–53) who also stress an organizational environment approach to be able to analyze inter-organizational networks. The second phase of organizational research since the 1990s shows an increased interest in groups of IOs and how IOs collaborate and compete with each other over influence or resources (for an argument for inter-organizational research, see Jönsson 1986). The approaches of network analysis and resource exchange thus show that organizational theories are well suited to be applied in IR research (Dingwerth et al. 2009, p. 26). In a similar vein, Hanrieder suggests incorporating trans-organizational coalitions into the research on IOs. She proclaims that coalition-building between different IOs also occurs on the sub-level, i.e., among departments, sections or divisions, which results in IOs being "open systems" (Hanrieder 2014b, p. 235).

By taking these trans-organizational communication paths into consideration, organizational research is able to tackle questions concerning intra-organizational cooperation and conflict more in-depth. Hanrieder calls these interlinkages between IOs 'boundary activities', and her organizational 'lens' makes it possible, first, to take different actors inside an IO into consideration, and second, to exhibit the parts of an IO that might contrast with how the IO wishes to be seen by the outside world (Hanrieder 2014b, p. 227). This understanding of IOs arises from the assumption that there are various perspectives and diverse power relationships inside an IO that need to be considered when interpreting organizational decision-making. By the same token, the organizational approach of 'new institutionalism' is developed further by DiMaggio and Powell (1991), as they focus on the broader context in which IOs operate. In this regard, the organizational environment and the interactions between different IOs are outlined by Powell and DiMaggio and are advanced into a comprehensive organizational approach (DiMaggio and Powell 1991). Following Reinalda and Verbeek, the external environment of an IO and the international regimes in which the IO interacts might thus develop and change, which ultimately impacts the working of the IO and leads to a shift in the organization's baseline (Reinalda and Verbeek 2004, p. 233).

The strand of organizational research that acknowledges the importance of the organizational environment and the broader context in which an IO operates bears resemblance to global governance (Dingwerth et al. 2009, pp. 19–20). Regardless of whether they differentiate between intergovernmental, transgovernmental and transnational governance (Dingwerth 2007) or broadly refer to global governance, scholars agree that (inter-) national decision-making means dealing with a variety of actors, one of which is organizations. Dingwerth and Pattberg (2009) and Powell and DiMaggio (DiMaggio and Powell 1991) focus on the polity context of organizational fields; the latter choose the transnational rule-making organizations in global sustainability politics as an exemplary organizational field. As such, IOs are ascribed a role in producing regimes in the international sphere that also influences inter-state relations (Dingwerth et al. 2009).[2] Other representatives of the global governance argument in organizational research highlight the importance of organizational action in comparison to organizational identity. By this means, the role of IOs in "governing at the transnational level" (Neumann and Sending 2013, pp. 132–133) is accentuated. Neumann and Sending criticize Barnett and Finnemore's (2004) understanding of the role of IOs in global governance, and they stress that it is important to acknowledge that IOs operate in a sovereignty-based international system that ultimately affects how powerful they are (Neumann and Sending 2013, p. 136).

To an even greater extent, research acknowledges that IOs do not act in isolation, but are surrounded by other actors with which "IOs share their bureaucratic and liberal features" (Neumann and Sending 2013, p. 136). Therefore, some authors advocate combining the autonomy-centered approach of the actor metaphor with the

[2]Although these authors attribute a certain regime-producing characteristic to organizations, this strand of literature is nevertheless rather critical as concerns the applicability of regime theory. Abbott and Snidal (1998) argue that regime theory deals with institutions and organizations in a rather distanced manner and thus has little to say on organizational structures that might influence or explain why states cooperate with each other (Abbott and Snidal 1998, p. 6). Regime theory's focus on IOs thus mainly centers on the way organizations help to reduce the transaction costs of cooperation between states.

principal-agent approach, which incorporates states' control of IOs (Neumann and Sending 2013, p. 137). Additionally, IOs interact with private actors in their respective organizational environments, and this networking activity possibly influences internal and external organizational factors. IOs might strive even more to be regarded in a better light, and therefore, interact and lean on private actors and institutions (Abbott and Snidal 2010, p. 316). Abbot et al. call this an 'orchestration' that postulates that IOs "have some degree of autonomy (or else a mandate from member states), an interest in orchestration, and the capacity to coordinate private actors and schemes" (Abbott and Snidal 2010, p. 324). An orchestration implies that an IO uses an intermediary actor in order to push an issue that is of major importance to the organization (Abbott et al. 2015, p. 4). That said, an IO's interaction with its international environment also guarantees the organization an unlimited choice of intermediary actors to approach its desired target actors.

Knill and Bauer (2007) differentiate between two relevant external factors that have an influence on management reforms inside organizations, the effects of the organizational domain and the effects of a changing constituency: The authors claim that an organizational environment has a major impact on whether a researcher bears witness to management reforms or not (Knill and Bauer 2007, p. 191). In this context, they argue that supranational organizations are more autonomous from their member states than IOs. If a member state of an IO changes its preference, the IO will thus react quicker to these changes, as this type of organization is very reliant on states (Knill and Bauer 2007, pp. 194–195). Knill and Bauer also hypothesize about the nature of organizations in terms of their size and about the question of whether a smaller IO is invariably more responsive to a changing organizational environment (Knill and Bauer 2007, pp. 194–195).

Overall, the interaction between an IO and its environment is always reciprocal. Although the environment largely influences an organization's action and sphere of influence, the IO also has an impact on the environment. In addition to the expert authority that many IOs possess, they also exert moral authority relating to their external environment (Barnett and Finnemore 2005, p. 172).

### 3.3.2 *The 'Member State' Approach*

One strand of organizational research centers on the question of the actorness of IOs: in other words, whether IOs constitute actors in their own right or whether these organizations are merely the executing organs of states. The so-called principal-agent approach acknowledges that organizations are agents with a particular self-interest that are mainly regulated and controlled by states (for a concise description of the state of research on principal-agent theory, see Graham 2014, pp. 368–369). According to Dingwerth et al., the origin of the principal-agent approach lies in the theory of firms (Dingwerth et al. 2009, pp. 18–19).

A major strand of organizational research has focused on the question of why states frequently use IOs as vehicles of cooperation, e.g. Abbott and Snidal (1998,

p. 4). Although the agent-centered research approach does not fully acknowledge the organizational environment, the principal-agent approach nevertheless allows a focus on questions concerning how great powers govern—within or through IOs (cf. Drezner 2007; Haas 1990, p. 58). Abbott and Snidal identify, first, centralization, and second, independence as the two functional characteristics that explain why states prefer IOs in comparison to other forms of institutionalization (Abbott and Snidal 1998, p. 4). In some cases, there are even different principals inside an IO that try to control the agent (the IO) (Reinalda and Verbeek 2004, p. 231). In the accounts Drezner outlines in *All Politics is Global* (2007), he ascribes an essential role to the member states of IOs in terms of agenda-setting and in terms of the way the IO will play a role in global governance. More specifically, changes in an IO's direction might be due to a changing coalition of member states, and thus, a new orientation of the IO (Reinalda and Verbeek 2004, p. 233). Then again, recent research looks into the communicative interaction between IOs and states and discusses the way that IOs constitute administrative authority beyond the member states (Benz 2017). In addition, research approaches IOs from the opposite angle and examines what factors influence the faithfulness of IOs to the mandates assigned to them by the member states (Graham 2014).

Moreover, organizational research does not only acknowledge the role of member states in terms of being policy brokers, but also analyzes the financial dependencies of IOs (Mehde 2007, p. 171). The monetary contributions of member states do not only influence an IO's orientation with regard to content, but are also interpreted by some scholars as improving the legitimacy of IOs (Mehde 2007, p. 171). In other words: "Only if the money is spent in the most efficient way possible can the organization be regarded as fulfilling the task assigned to it by its members" (Mehde 2007, p. 171).

Furthermore, the principal-agent model is helpful if organizational research aims at tackling questions concerning how member states delegate their authority to IOs in order to influence an IO's decisions. Nevertheless, Kleine (2013b) goes one step further and proclaims that the principal-agent model fails to incorporate the subcultures inside IOs. In essence, the concept of IOs as agents only sees IOs as unitary actors, as opposed to their principals, the member states. Therefore, it would be more accurate to acknowledge that IOs consist of various departments and sub-groups that "unofficially 'belong' to a state or a group of like-minded countries" (Kleine 2013b, pp. 321–322). As such, Kleine's approach adds another level of analysis to the traditional principal-agent model and provides insight into the national fiefdom of the 'agent'. While making it possible to analyze the complex daily work of IOs, the theory of national fiefdom also implies that the member states exercise more control over IOs than acknowledged by the principal-agent model (Kleine 2013b, pp. 322–323). In a similar vein, Haas talks about 'superpowers' inside IOs that have a hegemonic role due to their financial contributions and voting powers (Haas 1990, pp. 6–7). These—most likely—Western states bring Western beliefs and Western modes of problem-solving to IOs (Haas 1990, p. 58).

When looking into the behavior of the member states inside an IO, it is thus very likely that one will discover "inner elites" (Haas 1990, p. 58) who are frequently called in as experts and consulted in the decision-making of the IO's committees. By networking and issue-linking, these elites possibly exercise power over all organizational matters, and thus, need to be identified for a comprehensive organizational analysis.[3] Haas' (1990) and Kleine's (2013b) approaches thus contribute to the literature on informal governance within IOs. This is a strand of research that continues to receive less academic attention than formal political procedures (Kleine 2013b, p. 321).

### 3.3.3 The 'Character of the Bureaucracy' Approach

In response to the principal-agent approach mentioned above, another strand of organizational research centers on IOs as bureaucracies. From this perspective, IOs are not merely understood as agents of the principal's wishes, but rather as complex bureaucratic constructs on their own. The approaches that fall under the bureaucratic strand aim at opening the 'black boxes' of IOs. In this section, the bureaucracy approach is condensed into its main predications: first, theoretical assumptions that focus on the bureaucracy of IOs in general; second, theories that make a statement on the workforce of IOs; and, third, assumptions that point to the role of leadership in IOs (see Fig. 3.1).

First, the theoretical basis for the 'character of the bureaucracy' approach is laid out by March and Simon, who demonstrate that all decision-making and organizational change in IOs has to be interpreted against the background of a scarcity in time and attention (March and Simon 1976). IOs are not able to include all possible alternatives and the ensuing consequences in their decision-making. Additionally, March notices that IOs will never be able to target all their short- and long-term goals simultaneously (March 1989, p. 3). In his pragmatic interpretation of IO behavior, March further proclaims that it must be taken into consideration that most of the decisions taken by IOs do not meet the intended target (March 1989, p. 3).

Barnett and Finnemore—with their path-breaking contributions in *Rules for the World* (2004)—can be assigned to this strand of organizational research. By concentrating on the autonomy of IOs, the authors argue that it is the administration of IOs that gives rise to IOs as independent and autonomous actors in the international sphere (Barnett and Finnemore 2004). Although Barnett and Finnemore stress an intra-organizational perspective, they still argue in the broader global governance context: IOs are seen as agenda-setters for state interests and for global governance (Barnett and Finnemore 2004). From a critical stance, IOs might thus provide for employment-creation measures by themselves: As the workforce of IOs believes in the need for more bureaucratic interventions, the bureaucrats will always push for a stronger and more influential IO (Hanrieder 2014b, p. 225). From this perspective, the culture of an IO can be understood as "represent[ing] the internal friction that

[3]For more details on issue-linkage, see Haas (1990, pp. 76–80).

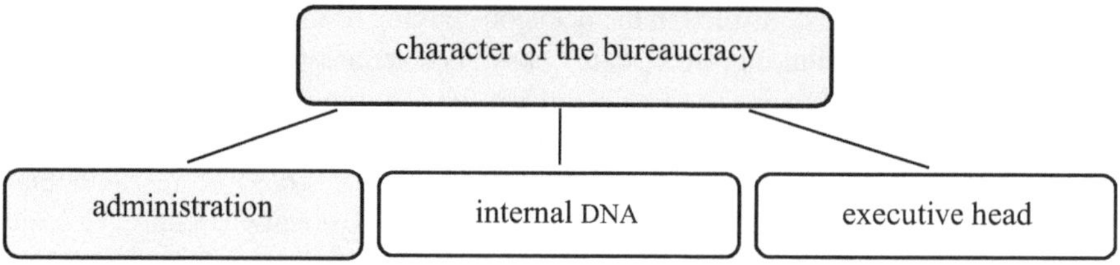


**Fig. 3.1** Depiction of the 'character of the bureaucracy' approach with a focus on the 'administration'

propels change in a slow, path-dependent direction that produces outcomes that are not necessarily congruent with the preferences of the organization's powerful shareholders or the intentions of internal reform initiators" (Weaver and Leiteritz 2005, p. 369). That said, the organizational cultures and the agenda of bureaucrats can change (see Sect. 4.2). In reference to the above-mentioned metaphors for IOs—arena and actor—Barnett and Finnemore refer to IOs as actors that are able to influence and operationalize policies for states (Barnett and Finnemore 2004, p. 156). Organizational structure and organizational culture thus correspond to the political environment of an IO, and the focus here lies on the question of "how IOs have authority over and thus autonomy from states" (Neumann and Sending 2013, p. 136). That is not to say that Barnett and Finnemore omit the role of states completely; rather, they acknowledge that in the end, it is states that set up IOs and keep them alive (Neumann and Sending 2013, p. 136).

In fact, this strand of research also advocates focusing on the internal composition of IOs, as this makes it possible to take the different arenas inside an organization into consideration. In this context, Haas discusses epistemic communities inside IOs; by this, he means "invisible colleges, networks of the like-minded not employed in the same university, laboratory, or think tank" (Haas 1990, p. 42). The relevance of these internal arenas lies in their hidden influence in committees, advisory groups or certain organs in the organization (Haas 1990, p. 42). Furthermore, Veit and Schlichte (2009) argue that an IO might dispose of various arenas, e.g. the headquarters, the base camps in member state countries, and the bush offices. The theoretical assumption of a lack of communication between the headquarters and the field offices is considered by Hanrieder in her analysis of the World Health Organization's (WHO) policies; she shows that the corporate program budget hardly affected the regional budgets, e.g. in the African offices (Hanrieder 2015a).

Second, the workforce of IOs and the internal DNA of organizations are receiving more and more academic attention (see Fig. 3.2). The decision-making inside IOs was originally described by referring to the personality traits or characteristics of individuals (March 1989, p. 5). Therefore, it was believed that an IO possesses an identity and a character with a specific set of individual preferences similar to a single individual. Most strands of organizational research nevertheless assume that IO decision-making is influenced by multiple actors, and they thereby contradict the interpretation that there is only one set of preferences inside an IO (March 1989,

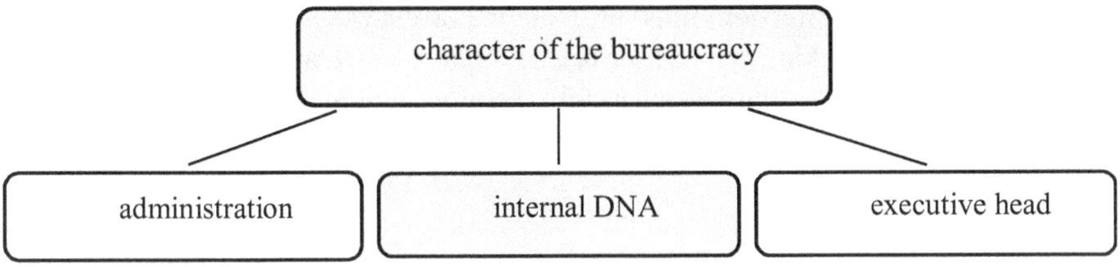


**Fig. 3.2** Depiction of the 'character of the bureaucracy' approach with a focus on the 'internal DNA'

p. 5). The knowledge production inside and outside of an IO is thus filtered by the epistemic community surrounding the IO (Haas 1990, p. 40). Furthermore, the secretariats, as well as the overall staff of IOs, are all highly professionalized in their respective fields, and thus introduce the respective organization to the knowledge base of their specific discipline, e.g. medicine, law, natural sciences, etc., with the result that "these professionals often [. . .] act in concert with like-minded professionals not in the employ of the organizations, but linked to them through nongovernmental organizations having consultative status, or by way of service on advisory panels of experts" (Haas 1990, p. 40). Recent research also focuses on the secretariats of IOs in comparison to public administration at the national level (Knill et al. 2017b, pp. 176–177). It hereby shows that international administrative bodies dispose of high levels of autonomy as compared to their national counterparts and that the secretariats of IOs actively shape their organizational environment (Knill et al. 2017b, pp. 177–181).

Research on the power constellations within an organization has been part of organizational research since early on, although it is difficult to measure power in IOs. One way of defining power in IOs is by referring to them as internal structures in institutions, and in conjunction with that, differences in resources (March 1990, pp. 140–141). March (1989) is one of the first to focus on the clash of interests inside IOs. More specifically, March claims that IOs are entities that are mistakenly thought of as innocent and harmonious, whereas he thinks that the different actors inside an IO manipulate information and lie, as long as it fits their respective self-interests (March 1989, p. 6). Furthermore, recent research focuses on hierarchies inside IOs and the respective influence of the different hierarchical levels (Eckhard and Mattmüller 2017, p. 242). Concerning the hierarchies inside IOs, it is helpful to remember Luhmann's conception of organizations as being autonomous in purpose, hierarchy and membership (see Sect. 3.1). Hierarchies clearly demarcate the roles and social relations inside an organization (Luhmann 1976, pp. 161–162). As such, intra-organizational hierarchies coordinate how the individuals behave. When an individual inside an organization receives a stimulus, he or she follows with a certain reaction that is within an organization-specific repertoire of possible actions (March and Simon 1976, p. 37). In Luhmann's understanding, all employees and their day-to-day work need to be somehow related to the organizational goals and purposes (Luhmann 1976, pp. 109–110). According to this understanding, every task of an employee of an organization thus ultimately correlates with the organization's overall goals. This thinking can be ascribed to Taylor, who claimed that if a company has its goals sorted out, it only needs to find the most suitable individual to fulfill the job (Taylor 1919, pp. 45–46). Weber had a similar understanding of how organizations choose their employees, and he agreed with the order of first deciding on organizational goals and tasks, and then, choosing the appropriate employee for the task in a second step (Weber 2009, p. 12).

This strand of research also sheds light on the role of bureaucrats and the extent to which they exert influence on IOs. Two thirds of the IOs exist because of the combined effort of states and bureaucrats, and some researchers thus assume that if bureaucrats actively involve themselves in the creation of an IO, the organizational structure of an IO will ultimately reflect this—i.e. less control being exerted by states and more organizational autonomy (Johnson 2013, p. 183). Here, the focus thus lies on the design processes of IOs and the way bureaucrats influence this stage of an IO's life. Johnson argues that states often accept bureaucrats' involvement because they function as a buffer zone between differing national objectives (Johnson 2013, p. 188). Bureaucrats often have previous working experience in other IOs, which leads to the conclusion that IOs have a different design when bureaucrats help design them. This understanding of IOs is thus critical of a principal-agent interpretation; instead, the bureaucrat approach allows for a more in-depth analysis of the role of the agent and its strategic and self-interested behavior. In other words, this strand of research acknowledges that the principals, i.e. the states, cannot completely control IOs (Johnson 2013, p. 186).

An elaborated focus on bureaucrats' behavior in IOs also allows to consider the diverse objectives that guide organizational actions. That said, the staff of IOs "want to further their mandates, and their professional backgrounds and expertise shape their interpretation of their mandates and the appropriate means to pursue them" (Barnett and Coleman 2005, p. 597). These objectives might differ from those of states, and an IO's organizational design might even be geared toward impeding state control (Johnson 2013, p. 186). In terms of material security, this strand of research ultimately touches upon questions of organizational design, and thus, helps to analyze the funding and resource structure. For this reason, bureaucrats try to diversify their IOs' funding schemes by relying on resources from states, as well as from external actors. Similarly, organizational legitimacy is central for bureaucrats, but it might be confined by states' control and persuasion. If states can easily influence IOs' strategies and decision-making, it will be hard for the respective IOs to argue that they represent more voices than just those of the biggest donors or strongest member states (Johnson 2013, p. 186). In the analysis of an IO's degree of legitimacy, it is important to examine the organizational design, veto power mechanisms and resource monopolization. In essence, insulating itself from state control and state interference is pivotal for an IO. According to Johnson, the essential factors for measuring state control—the dependent variables—are state financing, oversight meetings, unilateral vetoes, and government representatives (Johnson 2013, pp. 189–190). In conclusion, the strand of literature on bureaucrats forecasts a situation in which an IO that was created with the help of bureaucrats is characterized by few mechanisms of state control, whereas an IO that was created by states has more mechanisms of state control in its organizational design (Johnson 2013, p. 188).

As suggested by Barnett and Finnemore, the "divisions and subunits within the organization may develop their own cognitive frameworks that are consistent with but still distinct from the larger organization" (Barnett and Finnemore 1999, p. 719). Once again, the organizational culture is an important factor to understand what is perceived as good or as bad behavior in the organization (Barnett and Finnemore 1999, pp. 699–702). For this reason, the communication flows inside a bureaucracy become relevant. One theoretical assumption is that the superior controls the flow of information in an IO, and he or she decides who receives inside knowledge and who does not (Kühl 2011, p. 79). In reality, the internal communication goes through its own channels and cannot be steered in a top-down manner. Rather, informal ways of communication develop over time and grow, little by little, into their own communication network (Kühl 2011, p. 79).

Finally, the research on the internal DNA of IOs makes important assumptions about the recruitment and organizational practices of IOs. Recruitment processes inside IOs are built around the principle of agreement (Brunsson 2003, pp. 14–15). By means of only recruiting individuals who fit into the organizational culture of an IO, the 'organizational DNA' is preserved. According to Brunsson, the daily operations of an IO also rigorously function on the basis of agreement (Brunsson 2003, pp. 14–15). In a similar vein, Ege and Bauer shed light on the internal structures of IOs, the impact of hierarchical organizational behavior and well-established rules for decision-making (Ege and Bauer 2013, p. 143). Ege and Bauer (2013) claim that it is helpful to approach an IO by means of distinguishing between the political and administrative branches: The former thus includes focusing on the assembly meetings and the member state representatives, whereas the latter sheds light on the hidden coordination, preparation and policy implementation processes undertaken by the bureaucracy (Ege and Bauer 2013, p. 143).

Third, this theoretical approach also focuses on the characteristics of the executive head (see Fig. 3.3). This research strand thus falls into the broader field of leadership research. It is challenging to define leadership because it is a concept that is based on a highly subjective interpretation and is also "strongly influenced by one's philosophical beliefs on human nature" (Bolden et al. 2011, pp. 17–18). The

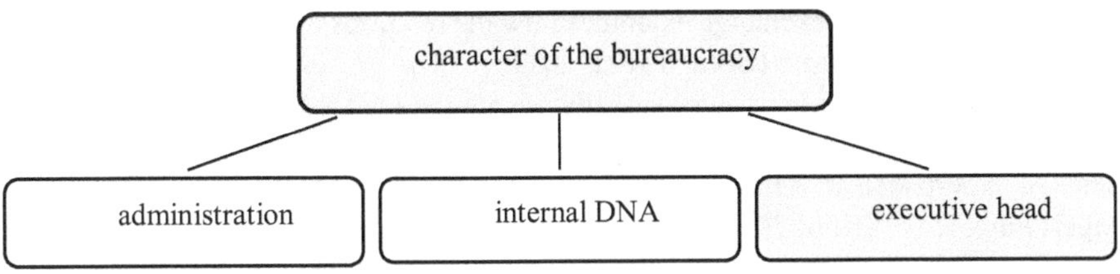


**Fig. 3.3** Depiction of the 'character of the bureaucracy' approach with a focus on the 'executive head'

origins of the current leadership studies are often dated back to the research done in the United States during and after World War I on the question of how to recruit military personnel (Bolden et al. 2011, p. 26). Hence, the role of leaders is approached from different angles: Some approaches focus on the character traits of effective leaders, the various possible leadership styles or the question of how external factors influence how a person leads.[4]

This thesis takes an interest in the executive head of an IO and is therefore guided by the literature on executive heads. Although the theoretical approaches on executive heads all stress different elements and characteristics, the literature overall differentiates between Cox's three major characteristics: legal-institutional, idiosyncratic (personality and leadership style), and ethical-normative (Cox 1969). Whereas the legal-institutional characteristic highlights the constitutional powers of the executive head, the idiosyncratic characteristic stresses the character traits of the executive head (e.g. charismatic vs. bureaucratic). The ethical-normative character of an executive head emphasizes the international role of the individual and what this international character implies for the role of the executive head. In the context of organizational theory, the assumption is that individual leadership impacts the decision-making inside an IO, and thus, the way the organization acts in the broader organizational environment (Koremenos 2005, p. 36). By the same token, Nye discusses the question 'Do leaders matter?', and he adds to the academic discussion as regards the relationship between leadership and power (Nye 2008). Generally, the leader of an organization often sees him- or herself confronted with being made responsible for organizational failures, although the leadership is not necessarily the cause of a certain organizational behavior. Whereas some might define a leader as "someone who helps a group create and achieve shared goals" (Nye 2008, p. 18), others argue that the ideal leader should "understand the strategic structure of their organization and enact the appropriate, context-specific strategies" (Koremenos 2005, p. 66).

Thereby, the theoretical assumption is that a superior's role does not suffice if it always aims at perfect harmony. In other words, the superior can make decisions without deliberating about whether a subordinate agrees (Luhmann 1976, pp. 96–97). Instead, the outcome needs to satisfy the customers' wishes, as well as the employees' demands. Research shows that employees accept and understand the need to change a well-established organizational culture only to a certain extent (Kühl 2011, p. 73); in these circumstances, the superior thus has to fall back into an authoritarian leadership style and simply push the organizational decisions. Research on the leader-member exchange examines how the relationship between the different actors develops over time (Bolden et al. 2011, p. 30).

A common mistake is to automatically equate the leadership in IOs with leadership in expert knowledge. But due to the specializations and the huge number of fields of expertise in an IO, it is seldom the superior who possesses the most in-depth expert knowledge (Kühl 2011, p. 77). Chances are that the hierarchical authority in

[4]For more details on the different leadership approaches, see Bolden et al. (2011).

an IO thus differs from the professional hierarchy. In the daily routine, it might be necessary to spare the superior all the details because the person in leadership is only able to care about a certain amount of information in a specific time span. That said, the subordinates are entrusted with deciding what informational bits are important and relevant enough to be passed on to the superior (Kühl 2011, p. 81). As a result, the decisions made by the subordinates significantly influence the decision-making of the superiors, and the origin of organizational reforms or changes might thus also spring from below the management and leadership level (Kühl 2011, p. 81). Luhmann argues that in some cases, it might be better to deliberately keep a superior's information flow low or to systematically cut him or her off from the information altogether (Luhmann 1976, p. 117). The theoretical concepts around the executive head of an organization thus need to be interpreted against the background of broader organizational research that also takes a subordinate's influence into consideration. Executive heads are not per se the most powerful actors of the IO. Rather, the leadership and management department is merely the one under the most adaptive pressure and also confronted with empowered subunits (Hanrieder 2015a, p. 221). Nevertheless, hierarchical structures inside organizations exist and should be taken into consideration in the evaluation of organizational decision-making and operations.

Following this depiction of research on the character of the bureaucracy, the next section outlines the intra-organizational strategies. This thesis separates the approaches in order to have analytical clarity, even though the 'character of the bureaucracy' and the 'intra-organizational strategies' presuppose each other and often overlap. Despite these interlinkages, the two theoretical 'lenses' still take a different look at organizations as will be shown in the following section.

### 3.3.4 The 'Intra-Organizational Strategies' Approach

Research on the internal composition of IOs also implies asking critical questions concerning organizational behaviour (see Fig. 3.4). This is exemplified by—amongst others—Brunsson's (2003) focus on hypocrisy in IOs.[5] Brunsson explores organizational rituals that solely serve the purpose of putting forward a certain image to the outside world (Brunsson 2003, p. 7). The recipients of an organization's propaganda machinery are the other actors surrounding the organization as well as the intra-organizational audience. According to Brunsson, it constitutes a far-reaching failure if organizational research does not remove an organization's mask when interpreting its actions.[6]

[5]More specifically, Brunsson aims to "analyze ways in which organizations handle inconsistent norms and how this affects the rest of their organizational behavior" (Brunsson 2003, p. 9).

[6]At the same time, Brunsson (2003) underlines the need of organizations to be perceived as legitimate by their constituents and by their environment. Here, an organization achieves legitimacy if the resources provided by an organization reflect what the environment expects in terms of organizational action (Brunsson 2003, pp. 14–15).

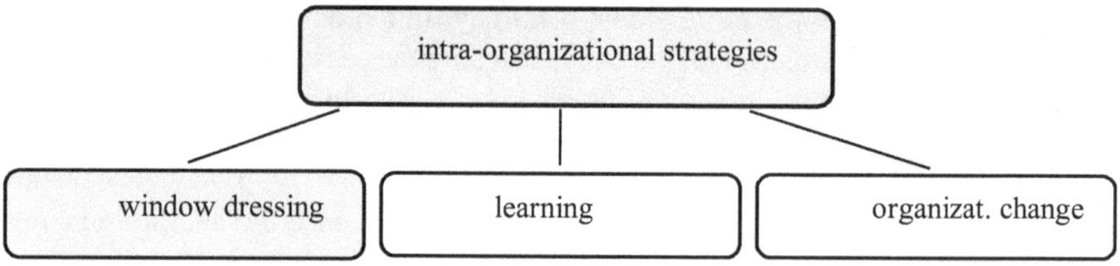


**Fig. 3.4** Depiction of the 'intra-organizational strategies' approach with a focus on 'window dressing'

Kühl (2011) stresses the difficulty that researchers or journalists face in asking about the intra-organizational state of affairs. The reactions are embellished stories about harmonious, democratic, and thus, legitimate decision-making inside an IO that almost constitute a conspiracy of silence (Kühl 2011, pp. 136–137). That said, organizational research always has to grapple with wishful thinking versus the bare truth on the part of an organization's insiders. This particularly applies to the impression one sometimes gets that IOs are very aware of the ravages of time, and they thus frame their own behavior and values accordingly. The challenge in organizational research is to uncover the "second realities" (Kühl 2011, pp. 137–138) that are deliberately invented for the outside world and have little in common with what is going on inside. These socially constructed façades of IOs need to be repeatedly revived and suited to the realities of the respective times (Luhmann 1976, p. 112). Because organizations fear that they might reveal certain paradoxes to the outside world, they rather only present one part of reality, and thus, tend to idealize their organization (Luhmann 1976, p. 113).

According to research, the aims of an organization's window dressing are manifold. First, the organization uses a façade in order to present a certain image of itself to the outside world. The organization's constituents copy and recopy this behavior by means of specific intra-organizational languages, symbols and daily practices (Kühl 2011, pp. 138–139). This mainstreaming inside an organization ultimately helps put forth a uniform image to external actors such as states, other IOs, the media and the general public. Organizational research thus aims to expose the stories behind the carefully reprocessed organizational corporate design, websites and propaganda. More specifically, this strand of organizational research intends to uncover what kind of image an organization wants to sell with a certain appearance (Kühl 2011, pp. 138–139). Second, by means of window dressing, an organization protects its internal actions. A certain amount of closed and secretive behavior is necessary for intra-organizational decision-making to function properly. If all processes, discussions and communication flows are presented in real time, the decision-making will suffer from the external involvement of pressure groups and member state interests (Kühl 2011, p. 140). Third, the window dressing of IOs aims at not being recognized as such. In other words, the façades try to hide an organization's use of a façade (Luhmann 1976, pp. 112–113).

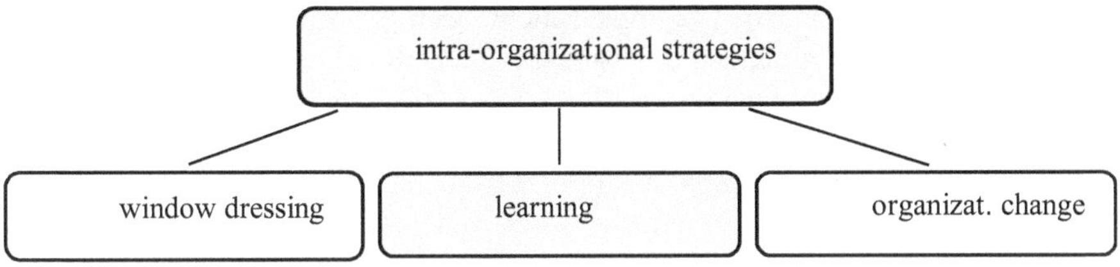


**Fig. 3.5** Depiction of the 'intra-organizational strategies' approach with a focus on 'learning'

This strand of research interprets IOs as icebergs (Kühl 2011, pp. 161–163): The observer is able to see the top of the iceberg, where everything is transparent and visible. Additionally, organizational research assumes that there is a hidden part of the iceberg. Here, there are inner organizational dynamics, ideas of the workforce and informal practices that are not visible to the external observer. The organizational behavior implies that although the superior asks subordinates to communicate openly, the management does not know what is hidden from them and how the staff decided on which information is formal and which is informal (Luhmann 1976, p. 117). What this strand of organizational research thus aims at is to explore both parts of the iceberg, as well as to understand the interplay between the top and the bottom half of IOs (Kühl 2011, pp. 161–163). Therefore, this theoretical 'lens' claims that in order for an IO to start a reform process, it needs a window of opportunity for reform, as well as an actor inside the organization who is willing to set the agenda for change (Kerler 2007, p. 136).

In a similar vein, and building on the work of Haas (1990), the bureaucracy of IOs becomes important, due to the fact that organizations internalize learning patterns (see Fig. 3.5). The understanding of learning goes back to the concept of the learning cycle. As described by March (1989), the cycle starts with the assumption that individuals' preferences affect their actions, and these individual actions ultimately affect the choices made inside an organization. The internal choices made by an IO, on the other hand, influence the organizational environment, and this environment ultimately has an influence on individual choices (March 1989, p. 14). Nevertheless, March points out that the "ambiguity of history" (March 1989, p. 13) has a strong effect upon the learning cycle, and it is thus questionable if the cycle as such can ever be regarded as fully complete. Additionally, the learning of organizations involves questions of interpretation: The interpretation of relevant contexts is often the daily business of IO employees, and the decision-making should be interpreted as being highly contextual and based on myth and rituals (March 1989, p. 13). For Haas, the learning in organizations leads staff members to question earlier beliefs about "the appropriateness of ends of action and to think about the selection of new ones, to 'revalue' themselves" (Haas 1990, pp. 24–25). That said, research on organizational learning focuses on the criteria inside an organization's decision-making process that explain how policies are realized. This strand of research argues that most of the organizational decisions are the result of rule-following and are thus in line with what is perceived as appropriate and in accordance with the organization's overall agenda (March 1989, p. 16). The organizational learning processes also depend on the organization's structure, and this largely influences the extent to which individuals learn inside an organization.

In the current state of learning research, the IO is seen as either having a flat or an authoritarian organizational structure (Dreiack 2016, pp. 109–112). Some of the questions that this strand of research asks thus center on whether the IO possesses formal or informal characteristics in its structure, and the extent to which these structural criteria influence the connection between the individual and the IO (Dreiack 2016, p. 112). Learning research claims that the internal make-up is highly dependent upon what an IO's employees "have learned as appropriate in a particular situation or as part of a particular role" (March 1989, p. 8). When looking at organizational changes or new rules for decision-making in IOs, it is thus important to incorporate intra-organizational processes of norm creation. That said, organizational research assumes that the current rules and procedures are the result of a long process of optimization and learning; to rate whether an IO is effective or not thus comes down to the question of whether the IO is able to learn quickly and precisely (March 1989, p. 9). Still, the linkage between a problem and its solution is very dependent upon whether there is a simultaneity of events. This is exemplified by the so-called garbage can model, in which decision-making inside IOs is approached from a standard means-end frame (March 1989, p. 13). That said, the people inside IOs and the organization itself learn from past experiences in a very rational process (March and Olsen 1975, p. 356). Further, if one assumes that the employees receive appropriate information and that they draw the right conclusions, this learning process inside an IO will likely lead to positive results and improvement.

As recent examples of the advancing research in this theoretical strand, Hall looks into the learning processes of IOs inside a specific regime—the climate change regime—and thereupon conducts inter-organizational comparisons between different IOs in this regime, the IOM, the WHO, and the UNHCR (Hall 2015, p. 79). Dreiack differentiates between a threefold of learning processes inside IOs: first, the learning of individuals inside IOs, second, the learning of the collective inside IOs, and third, the learning of the specific organs inside IOs, such as the organization's secretariat, etc. (Dreiack 2016, p. 108). Therefore, the research on organizational learning starts from the assumption that an IO's environment is as complex as the organization itself and that decisions that might seem intelligent and future-oriented might appear less positive later on (March 1989, p. 16).

The research on organizational change in IOs (e.g. DiMaggio and Powell 1991) and the research on discrepancies between mandates and organizational behavior (e.g. Hanrieder 2014a; Starke 2013) discuss different mechanisms (see Fig. 3.6).[7] The literature on static IOs, especially, has been challenged by recent constructivist approaches. IOs are not immune to their constituents' yearning for change and

[7] As representatives of this research strand, Jetschke and Theiner (2016) differentiate between five mechanisms that indicate a distinct channel of influence on organizational change: learning, peer-based learning, legitimacy-driven mimicry, competition-driven mimicry, and soft coercion.

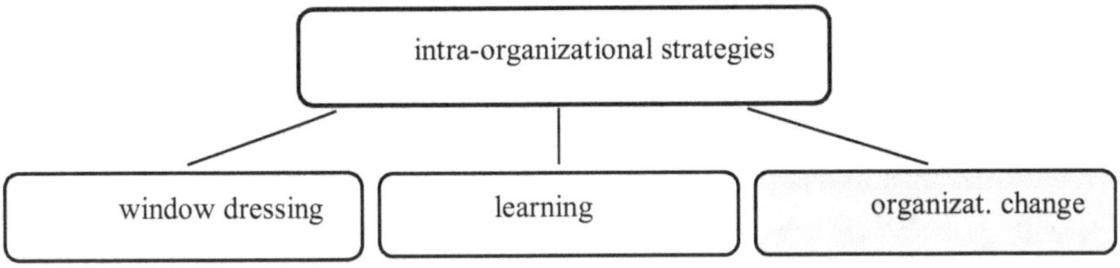


**Fig. 3.6** Depiction of the 'intra-organizational strategies' approach with a focus on 'organizational change'

renovation, and these internal struggles might very well lead to change in an IO (Hanrieder 2014b, p. 225). Nevertheless, for IOs to change their time-tested programs, they need some kind of outside pressure (Kühl 2011, p. 155). Further, even if changes inside an organization are recognizable, they often just modify some elements of an IO's agenda instead of changing the overall organizational goals.

When explaining organizational change, scholars often lean on institutionalism. Fioretos demarcates the field of research that focuses on organizational change into, first, studies based on rational choice institutionalism and state-centric IR research; second, sociological institutionalism approaches; and, third, studies based on historical institutionalism (Fioretos 2011, p. 368). Explanations of organizational reform from a rational institutionalist standpoint start from the assumption that states only join an IO when their participation satisfies their goals (Hanrieder 2015a, p. 218). Consequently, rational institutionalism expects organizational change in those cases in which member states no longer see themselves represented in an IO's direction or agenda (Koremenos et al. 2001); in other words, "states pay careful attention to institutional design" (Koremenos et al. 2001, p. 762). Sociological institutionalism, on the other hand, considers the normative orders and the structural qualities of institutional environments to be the most relevant. This theoretical 'lens' has the additional value of tackling the question of "how institutional legacies affect the degree to which power resources can be harnessed" (Fioretos 2011, p. 370). Adding to this, historical institutionalism looks into the possibility of an organization incrementally changing over time.

One rather structuralist strand of historical institutionalism focuses on the importance of exogenous factors and path-dependency, whereas a second, agent-centered direction stresses the role of powerful actors that have the capacity to change policy settings (van der Heijden 2012, p. 60). According to Hanrieder, a historical institutionalist lens allows one to acknowledge historical developments and struggles as being the origin of an organization's structure. As such, IOs "become themselves determinants of their future evolution" (Hanrieder 2015a, p. 218) and do not merely reflect the power struggles of their member states. The advantages of historical institutionalism thus lie in the theory's ability to focus on the complex institutional dynamics that trace organizational changes back to the clash of existing privileges inside IOs and external pressure on the organization (Hanrieder 2014a, p. 331). Fioretos (2011) adds that historical institutionalism helps the field of IR to mature and to include questions of change and continuity in global governance. Also, historical institutionalism sheds light on the founding moments of organizations, and the theory regards these as important markers for the long-term power relations of an IO (Fioretos 2011, p. 369).[8] This strand of theory thus helps to explain why certain ideas arise in IOs that have political consequences in the long run (Fioretos 2011, p. 375). Historical institutionalism also propagates the concept of critical junctures.[9]

The above-mentioned agent-centered strand of historical institutionalism focuses on different kinds of organizational transformations, such as layering, conversion, drift, displacement and exhaustion (van der Heijden 2012, p. 61). Another strand of historical institutionalism is the so-called punctuated evolution lens. Here, the emphasis lies not only on institutional path-dependency but also on ideational path-dependency, as ideas and norms are taken seriously (van der Heijden 2012, p. 61). According to the punctuated equilibrium understanding, crises are considered as triggers of transformation. By means of a crisis narrative, the parties involved direct the future policy transformation. In comparison to historical institutionalism, gradual change concepts assume that the moments in which an organization—or rather its workforce—experiences "choice points" (Capoccia and Kelemen 2007, p. 350) are fairly regular. Here, the elements where organizational reform might originate are an organization's chairs, voting rules, executive boards and leadership selection (Bradford and Linn 2007, pp. 5–7).

Summing up, this thesis' understanding of 'organizational change' implies that there have been incidents that paved the way to the UNHCR's interest in EDPs and

---

[8]That said, historical institutionalism can be understood as a theoretical tradition and not as a general theory as such; it focuses on well-established presumptions, such as the claim that the more time passes after an event or a decision, the more difficult it is to reverse it afterwards and that accidental events can have long-term effects (Fioretos 2011, pp. 370–371).

[9]This implies that during an organization's lifetime, which is mainly characterized by stability and routine, there are also times of dramatic change. Through the lenses of historical institutionalism, the decisions made during these critical phases have a lasting impact on the lifespan of an organization and influence the strategies on which an organization embarks in the future. During these short periods, the normal organizational structures are relaxed, and political actors can use their power to influence organizational choices. The choices made during a critical juncture potentially have a game-changing character for an organization, although it is unclear what consequences these decisions will bring: "Contingency, in other words, becomes paramount" (Capoccia and Kelemen 2007, p. 343). Capoccia and Kelemen stress that it is important to differentiate critical junctures from incremental processes in an IO, although both can lead to transformational change inside the organization (Capoccia and Kelemen 2007, p. 368). Additionally, a brief critical juncture has a higher chance of having a lasting impact because the organizational structures and constraints are not able to re-emerge quickly enough to constrain the actors. In summary, "a critical juncture should have a duration that is short relative to the path-dependent process it initiates" (Capoccia and Kelemen 2007, pp. 350–351), and in critical junctures, the element of contingency thus constitutes the determinant. Further, the outcome of such a critical juncture is not necessarily change; rather, it might very well lead back to the equilibrium because change was ultimately rejected as a sought-after option (Capoccia and Kelemen 2007, p. 352).

that one can identify long-ranging and distinct changes inside the organization's bureaucracy due to these initial incidents.

## 3.4 A Critical Discussion of the State of the Art

Building on the above-mentioned current state of 'research on IOs' and the current state of research on 'organizational research', a few theoretical and empirical aspects will be discussed. The theoretical expandability hereby centers on, first, theoretical pluralism in organizational research, second, temporality in organizational research and, third, the element of power in organizational research. Concerning the empirical research, this section subsumes four broad research areas: first, research on the EU, second, research on the UN system, third, research on a group of other IOs and, fourth, research on the administration bodies of IOs. The depiction of the empirical research shows that research on the UNHCR is still scarce. Even though most renowned researchers have stopped treating IOs like 'black boxes', this thesis conforms with the previously declared empirical and theoretical research gaps (e.g. Dijkzeul and Beigbeder 2003, p. 11) and shows that not all of these gaps have been satisfactorily approached yet.

### *3.4.1 Theory-Centered Reflection*

First, the organizational research that reaches across theoretical boundaries and applies *theoretical pluralism* is still scarce. As the behaviors of IOs are perceived as merely responses to the interests of powerful states, IR research does not expect undesirable behavior from IOs (Barnett and Finnemore 1999, pp. 715–716). Further, in public administration research, IOs are also rather marginalized, especially when compared to nonprofit or public organizations (Dijkzeul and Beigbeder 2003, p. 10). Therefore, during the last couple of years, researchers have called for bridging this divide in the research (Barnett and Coleman 2005, p. 615; Nielson et al. 2006, p. 132). By means of opening the 'black boxes' (IOs), the behavior of IOs took center stage in research, as depicted by the approaches that uncover organizational dysfunction and pathology (Barnett and Finnemore 1999, p. 716). In a similar vein, Hanrieder advocates a cessation in the division between the rational choice approaches (for cases of the adaptation and design of IOs) and constructivist approaches (for cases of inertia and pathology in organizational cultures) (Hanrieder 2015a, p. 218). Approaches of theoretical pluralism also become apparent in research on organizational change. As an example, van der Heijden has called for theoretical pluralism in institutional change theories. He combines three institutionalist strands (van der Heijden 2012, p. 57) and also approximates the concept of institutional layering through the theoretical lens of Kathleen Thelen (van der Heijden 2011, p. 9).

In answer to this first area of theoretical expandability, this thesis aims to surpass the broad separation between the IR approach to IOs and the public administration approach to IOs. Hence, the thesis builds upon recent discussions about and research on a public administration perspective on IOs (Bauer et al. 2017a). As Ege and Bauer (2013) put it, "IR usually is better in explaining why IOs are created, whereas Public Administration is better suited to analyzing the policy-making role of IOs and their bureaucracies in day-to-day politics" (Ege and Bauer 2013, p. 135). The work at hand builds a bridge between the two approaches (IR and public administration). Therefore, the thesis settles in this theoretical gap and thereby tackles the problem of the applicability of IR theory in the IO context.

Second, the review of the literature shows the necessity of enhancing organizational theory that focuses on the *modality of organizational changes*, e.g. factors such as the time and the length of gradual changes in IOs. Whereas it must be acknowledged that principal-agent theory is able to explain why IOs start their respective learning processes without being pressured by their principals, this theoretical basis still fails to explain the qualities and varying degrees of an organization's learning process (Siebenhüner 2008, p. 93). In the context of empirical cases that involve gradual change within IOs, scholars thus deem agency theory and historical institutionalism as appropriate for exploring the interference of states in organizational reform. Nevertheless, Hanrieder (2014a) holds that further in-depth research is required on institutional change in general and gradual change in particular. More specifically, the long-term dynamics of change in IOs and the effects that result from these organizational reform processes have not been explored.

Furthermore, the much-used concepts of the historical institutionalism and punctuated equilibrium literature are afflicted with a theoretical vagueness when it comes to the different modes of change (van der Heijden 2012, p. 61). Van der Heijden raises concerns about the ambiguity between regular change and major change, and between regular shifts and sudden shifts in an organization (van der Heijden 2011, p. 10). Thus, the question arises: What is the transformation that is needed before one can declare an organizational change? Van der Heijden incorporates Thelen's understanding of change and sees her approach—to understand change as *either* incremental or radical—as one way of solving the above-mentioned question concerning the modes of change (van der Heijden 2011, p. 12).

In response to this second theoretical gap in the current state of organizational research, this thesis argues that most of the research fails to take the discrepancy between organizational rhetoric and organizational action into consideration. Although there is research on crisis narratives and narratives of change—e.g. securitization Methmann et al. (2013); Trombetta (2014)—this research ignores the inner modus operandi of IOs that might cause this kind of rhetoric. By the same token, some approaches, which are closer to organizational research, include ideational and narrative aspects, but the theoretical frameworks are rather superficial with respect to the temporal dimension of this rhetoric (and whether and how it changes over time). As a result, this thesis builds upon the organization's own security references as a starting point to look into the modality of the intra-organizational changes.

The organization's rhetoric thus adds the factor of time to the analysis, as the ideational depictions signify one building block of this analysis. On the one hand, an exclusive focus on the organization's actions and policies do not shed light on the way the organization interacts with the outside world (e.g. the rhetoric); on the other hand, an exclusive focus on the organization's rhetoric does not open the 'black box' or illuminate the organization's behavior.

Third, scholars of organizational theory argue that it requires more theory on *who pulls the strings* in an IO, e.g. the shifts of power between actors. In other words, a research gap in organizational research exists on the interlinkage of power and IOs (Hanrieder 2015a, p. 218). Although historical institutionalism is able to grasp power struggles, research either examines the way external pressure pulls and pushes IOs or the extent to which power structures influence the internal workings of an IO. Here, the focus shifts to another research gap, namely the comparatively small body of research that acknowledges IOs as both principals and agents (Gutner 2005, p. 14). Principal-agent approaches are criticized for being dissatisfying when confronted with cases of multiple principles (Gutner 2005, p. 20). Principal-agent approaches are not able to grasp why and how the agent's interests, as opposed to the principal's, emerge (Siebenhüner 2008, p. 94). In a similar vein, Kleine (2013b, pp. 321–323) points to the lack of research on subcultures inside IOs, as most theoretical approaches do not acknowledge that IOs might be controlled from the inside. As outlined by Hanrieder (2014b, p. 221), current research on IOs also fails to take into consideration the mission shrinking that some IOs experience.

Concerning the literature on leadership and executive heads, this thesis conforms with the research gap depicted by Schroeder who argues that although there is widespread literature on leadership, only a "few scholars examine whether their findings travel to international organizations" (Schroeder 2014, p. 342). More specifically, those theories contribute little to the research on executive heads; their theoretical 'lens' does not expect organizational change to be triggered by strong leaders inside an IO (Schroeder 2014, p. 345). In answer to this research gap, a few scholars—among others, Adebajo (2007, p. 142); Johnstone (2003, p. 442); Johnstone (2007, pp. 127–128); Kennedy (2007, pp. 158–159)—have looked into the topic of influential individuals in IOs who create norms and thus impact international politics. Kille, who has done extensive work on the role of the UN Secretary-General, points out that the "moral and religious dimensions of the office" (Kille 2007, p. 1) have not yet been examined in a detailed and comparative manner. Thus, Kille (2007) claims that more comprehensive analysis is needed on executive heads.

In answer to this third area of theoretical expandability, the thesis thus approaches the research gap defined by Ege and Bauer, namely that the current state of organizational research still fails to shed light on the bearer of agency within IOs (Ege and Bauer 2013, pp. 142–143). According to the two researchers, most research cannot provide sufficient insight into the bureaucratic characteristics inside an IO and the way they interact with other political actors (Ege and Bauer 2013, pp. 142–143). In this context, Bauer and Ege carve out the differences between national and international public administrations and hereby approach the questions as regards autonomy and control at the international level (Bauer and Ege 2017, p. 14). The thesis picks up this thread and looks into the external and internal influences that might have restricted the UNHCR's behavior. Furthermore, Knill et al. add to the question of who pulls the strings in an IO by delivering insight into the administrative styles, that is, by differentiating between entrepreneurial and servant styles (Knill et al. 2017a, p. 61), whereas Benz et al. elaborate on how international public administrations "use [their] persuasive coordination patterns to extend their impact on national policy making without having the power or capacities to implement a policy or enforce national implementation" (Benz et al. 2017, p. 163). The thesis considers these current debates and findings in organizational research and thereby deems it necessary to approach the administrative style as well as the organizational culture of the UNHCR. Therefore, it also builds on the work of scholars who attribute importance to organizational culture (among others, Barnett and Finnemore 2004, p. 159; Nielson et al. 2006, p. 132; Weaver and Leiteritz 2005, p. 369), and those who focus on the role of the rhetoric of IOs (e.g. Dingwerth and Pattberg 2009, p. 713; Schimmelfennig 1997). Even though the international public administrations of IOs command a large amount of autonomy (as compared to their national counterparts) (Bauer et al. 2017b, p. 183), IOs in general are under considerable constraint of both internal and external power struggles. The thesis assumes that these power struggles are reflected in organizational behaviour and it therefore incorporates organizational research on autonomy and control in IOs as well as bureaucratic and leadership styles.

To sum up this theory-centered reflection, this thesis examines IOs in an all-encompassing manner, taking into account the internal and external factors that steer organizational behavior. As a scholar who has already tried to tackle IOs in this overarching manner, Schechter has analyzed IOs by focusing on systemic conditions, as well as organizational characteristics and personal characteristics (Schechter 1987, p. 214). In a similar, yet more elaborated way, this thesis builds on the observation that IOs try to balance their organizational autonomy with the external pressures (Barnett and Coleman 2005, p. 593). In the broader sense, this thesis thus makes a contribution to the academic dialogue across the disciplines of IR and public administration. Adding to this, the thesis profits from and builds on the previous institutional turn in IR (Fioretos 2011, pp. 367–368).

### 3.4.2 Empiricism-Centered Reflection

The empirical gaps in the research on IOs add to the above-mentioned research gaps in organizational theory. Whereas the latter concerns the matters of organizational creation, reform, culture, behavior, bureaucracy, etc., the former pays attention to IOs. Some scholars suggest that research has often neglected to use sociological or institutional theories on IOs, and they thus call for more empirical research on IOs that considers organizational theories (Dijkzeul and Beigbeder 2003, p. 7).

Additionally, scholars have criticized the lack of multidisciplinary, comparative and diverse (e.g. Northern and Southern perspectives on IOs) empirical research on IOs (cf. Dijkzeul and Beigbeder 2003, p. 312). Nevertheless, many researchers have contributed to a better understanding of IOs by means of extensive and in-depth empirical research. A condensed depiction of this research, including some examples of respective empirical studies, is considered below. First, research on the *EU* is quite elaborate and focuses, among other things, on questions of authority in the EU (McNamara 2010)—e.g. through regulatory networks (Blauberger and Rittberger 2015)—and takes EU norms and rhetoric into consideration (Schimmelfennig 2003). Furthermore, Schimmelfennig and Sedelmeier's research on the EU centers on the question under which conditions the EU's external governance is most effective, with a special focus on the transfer of EU rules to Central and Eastern Europe (Schimmelfennig and Sedelmeier 2005). In a similar vein, Knill (2001) analyzes the extent to which EU policies have an impact on national administrations, whereas Schön-Quinlivan and Bauer examine administrative reform in the European Commission (Bauer 2007, p. 53; Schön-Quinlivan 2007, p. 31). Kleine thoroughly focuses on the importance of informal norms in the EU and the way the member states strengthen domestic interests parallel to European integration (Kleine 2013a). Taking EU research one step further, Jetschke and Theiner compare the EU with ASEAN and examine their institutional similarities (Jetschke and Theiner 2016, p. 18).

Second, the *UN system* is an actor on which many researchers have focused. More specifically, the UN is a common empirical case study in research that explores the inner workings of an IO (Barnett 2003), as well as the way an IO might influence international politics or have a stake in global governance (Rittberger 2001). Among others, Barnett (2003) recapitulates the role of the UN in the Rwandan conflict during the 1990s. Van der Lugt and Dingwerth address one of the offshoots of the UN, the UNEP (van der Lugt and Dingwerth 2015), and Schechter focuses on the United Nations Educational, Scientific and Cultural Organization (UNESCO) and the UNDP in his research on leadership in IOs (Schechter 1987, pp. 197–198). Similarly, Finnemore chooses the UNESCO as an example of an IO that teaches norms to states (Finnemore 2015). Andresen and Rosendal (2009) and Ivanova (2009) critically discuss the UNEP's role in the context of the global environmental regime. In a comparative approach to the UN, the EU and the WTO, Dingwerth et al. (2011) analyze whether these organizations govern on the basis of democratic principles, and if this is not the case, how the democratic deficit can be resolved. Siebenhüner also uses a comparative approach and compares the learning processes of eight different organizations (to name a few: UNEP, UNFCCC, IOM and World Bank) (Siebenhüner 2008, p. 105).

Third, a few *other IOs* are repeatedly represented in the research on IOs. The ILO has been used as a case-study by Ege (2016) and Baccaro (2015), whereas Baumann et al. (2007, pp. 184–185) focus on the role of the executive head of the ILO. The inner workings of the WHO are central in the work by Ege (2016), and Hanrieder takes the question of 'WHO orchestrates?' very seriously in her research (Hanrieder 2015a, b). In a similar vein, Finkle and Crane (1976, pp. 367–368) examine the

WHO's organizational values. Also, Hanrieder and Kreuder-Sonnen (2014, p. 342) break down the WHO's emergency policy with the theory of securitization. The reform processes of the World Bank are also used as an example of an IO's organizational change, e.g. by (Nielson et al. 2006; Weaver and Leiteritz 2005; Weaver 2008). In the context of the World Bank's claim of good governance, Weaver contrasts the bank's rhetoric with the bank's policies (Weaver 2008). Gutner examines the gap between the mandate, i.e. the stated goals of the IO, and the performance, i.e., the implementation of the World Bank's policy-making, and he theoretically builds on agency theory (Gutner 2005, p. 32). Similarly, Kerler (2007, p. 146) analyzes the reform process of the World Bank by taking NGOs and member states into consideration, and Park (2009) looks at the World Bank in the context of global environmental governance.

Chwieroth (2015) is a representative of the researchers who focus on the International Monetary Fund (IMF). He examines the normative orientations that shape the IMF (Chwieroth 2008, 2015) and how the staff at the IMF exercises informal governance control over the IO's decision-making (Chwieroth 2012). Also, Barnett and Finnemore (2004) examine the power constellations inside the IMF. In his research, Gutner questions the role of the IMF as a moral and good global actor (Gutner 2010). Zangl has published extensively on the General Agreement on Tariffs and Trade (GATT) and the World Trade Organization (WTO), and he elaborates on the way these organizations have influenced national policy-making in, for example, the US and the EU (Zangl 2006). The opposing perspective, namely on the extent to which states influence the WTO's practices, is tackled by Elsig (2015). Drezner's recent contribution to the research on IOs has also added to the strand of research on the WTO and the IMF. He stands up for these actors' role in the global economy and acknowledges the value of their response mechanisms during the financial crisis (Drezner 2014). With respect to the International Organization for Standardization (ISO), Prakash and Potoski (2010) shed light on the ISO as a global governor on questions of international standardization. Further, Barnett and Coleman (2005, p. 593) examine organizational change at Interpol.

Lastly, the relationship between IOs and transnational actors, e.g. NGOs, businesses, social movements and foundations, has been of growing interest for academia.[10] Steffek introduces a theoretical framework to identify the collaboration between IOs and NGOs, taking into account the incentives and the obstacles that go along with this kind of cooperation (Steffek 2013). Tallberg et al. examine IOs using transnational actors as "policy experts, service providers, compliance watchdogs, and stakeholder representatives" (Tallberg et al. 2013, p. 1). In global environmental governance, for example, the relationship between the leading environmental IOs and certain private partners has been analyzed by Andonova (2009), Pattberg (2009) and Timmer (2009).

Fourth, researchers have chosen the *administration bodies* of IOs (Liese and Weinlich 2006) and secretariats—for example, of Amnesty International (Hopgood

[10]On the accountability and representativeness of NGOs, see Steffek and Hahn (2010).

2006)—as their empirical cases. Johnstone sheds light on the role of the UN Secretary-General, using security theory as an explanation (Johnstone 2003, pp. 446–447). Biermann and Siebenhüner's (2009) miscellany presents research by various authors on, among other things, the secretariats of the International Maritime Organization and the UNEP, the secretariat of the Global Environment Facility, the ozone secretariat within the UNEP, the climate secretariat, the secretariat of the Convention on Biological Diversity and the secretariat of the United Nations Convention to Combat Desertification (Biermann and Siebenhüner 2009). In a similar vein, Bauer (2007) focuses on the European Commission's administration as an empirical case study (Bauer 2007). Concerning the role of the leadership of IOs, Schroeder summarizes the extensive research on executive heads that has been conducted on, for example, the UN Secretary General and the executive heads of the UNHCR, UNICEF, UNCTAD, UNDP, UNEP, and the World Bank (Schroeder 2014, pp. 340–343). Similarly, Baumann et al. (2007, pp. 184–185), who refer to executive heads and other influential individuals as reform brokers, focus on the leaders inside the World Food Programme (WFP) and the Food and Agriculture Organization of the UN (FAO).

In response to the empirical research presented above, this thesis sheds light on the UNHCR, an IO that has so far received comparatively little scholarly attention. To cite some examples, Barnett and Finnemore (2004) focus on the changing norms inside the UNHCR and Loescher et al. (2012) analyze the role of the UNHCR in the twenty-first century and its role as a global institution in matters of refugee protection. Hall (2015) examines both the UNHCR and the IOM in the context of their recent involvement in the climate change regime, and Loescher and Milner (2011) critically reflect upon the role of the UNHCR in the global refugee regime (Loescher and Milner 2011). Betts (2011) analyzes the institutional, political and normative character of the global governance of migration (Betts 2011). That said, this thesis acknowledges the UNHCR's role as *the* global player in the refugee regime and expects more than just an IO that is the puppet of states. With reference to these theoretical and empirical research gaps, the next chapter presents the research design of this thesis and how the chosen approach for this analysis ties in with the research outlined above.

To conclude, this chapter has provided a concise definition of the term 'international organization' and it differentiated between organizational research and the research on IOs. After discussing the applicability of IR theory in the research on IOs, the focus first shifted to the research on IOs. Second, this chapter classified the current state of organizational research into four strands of research that accentuates, first, the organizational environment, second, the role of states in IOs, third, the character of the bureaucracy, and fourth, organizational strategies. The approaches on the organizational environment and on the role of states hereby depict the external factors, whereas the approaches on the bureaucracy and the organizational strategies focus on the internal factors. Lastly, this chapter has critically discussed the state of the art from a theoretical and an empirical point of view. By this means, this chapter made the theoretical provisions for the development of the research design in Chap. 4.

## References

Abbott KW, Snidal D (1998) Why states act through formal international organizations. J Confl Resolut 42:3–32

Abbott KW, Snidal D (2010) International regulation without international government: improving IO performance through orchestration. Rev Int Organ 5(3):315–344

Abbott KW, Genschel P, Snidal D, Zangl B (2015) Orchestration: global governance through intermediaries. In: Abbott KW, Genschel P, Snidal D, Zangl B (eds) International organizations as orchestrators. Cambridge University Press, Cambridge, UK, pp 3–36

Adebajo A (2007) Pope, pharaoh, or prophet? The secretary-general after the cold war. In: Chesterman S (ed) Secretary or general? The UN secretary-general in world politics. Cambridge University Press, Cambridge, UK, pp 139–157

Andonova LB (2009) International organizations as entrepreneurs of environmental partnerships. In: Biermann F, Siebenhüner B, Schreyögg A (eds) International organizations in global environmental governance. Routledge, New York, pp 195–222

Andresen S, Rosendal K (2009) The role of the United Nations environment programme in the coordination of multilateral environmental agreements. In: Biermann F, Siebenhüner B, Schreyögg A (eds) International organizations in global environmental governance. Routledge, New York, pp 133–150

Baccaro L (2015) Orchestration for the "social partners" only: internal constraints on the ILO. In: Abbott KW, Genschel P, Snidal D, Zangl B (eds) International organizations as orchestrators. Cambridge University Press, Cambridge, UK, pp 262–285

Barnett M (2003) Eyewitness to a genocide: the United Nations and Rwanda. Cornell University Press, New York

Barnett M, Coleman L (2005) Designing police: interpol and the study of change in international organizations. Int Stud Q 49:593–619

Barnett MN, Finnemore M (1999) The politics, power and pathologies of international organizations. Int Organ 53:699–732

Barnett M, Finnemore M (2004) Rules for the world: international organizations in global politics. Cornell University Press, New York

Barnett M, Finnemore M (2005) The power of liberal international organizations. In: Barnett M, Duvall R (eds) Power in global governance. Cambridge University Press, Cambridge, UK, pp 161–184

Bauer MW (2007) The politics of reforming the European Commission administration. In: Bauer MW, Knill C (eds) Management reforms in international organizations, vol 6. Nomos, Baden-Baden, pp 51–70

Bauer MW, Ege J (2017) A matter of will and action: the bureaucratic autonomy of international public administrations. In: Bauer MW, Knill C, Eckhard S (eds) International bureaucracy: challenges and lessons for public administration research. Palgrave Macmillan, London, pp 13–42

Bauer MW, Knill C (eds) (2007) Management reforms in international organizations. Nomos, Baden-Baden

Bauer MW, Knill C, Eckhard S (eds) (2017a) International bureaucracy: challenges and lessons for public administration research. Public sector organizations. Palgrave Macmillan, London

Bauer MW, Knill C, Eckhard S (2017b) International public administrations: a new type of bureaucracy? Lessons and challenges for public administration research. In: Bauer MW, Knill C, Eckhard S (eds) International bureaucracy: challenges and lessons for public administration research. Palgrave Macmillan, London, pp 179–198

Baumann SM, Hagel M, Kobler B (2007) Management in change - the reform broker concept. In: Bauer MW, Knill C (eds) Management reforms in international organizations, vol 6. Nomos, Baden-Baden, pp 175–190

Benz A (2017) Kommunikative Verwaltungsbeziehungen zwischen Staaten und internationalen Organisationen - Zur Konstitution von Verwaltungsherrschaft jenseits des Staates. dms-der moderne staat-Zeitschrift für Public Policy. Recht Manage 10:211–227
Benz A, Corcaci A, Doser JW (2017) Multilevel administration in international and national contexts. In: Bauer MW, Knill C, Eckhard S (eds) International bureaucracy: challenges and lessons for public administration research. Palgrave Macmillan, London, pp 151–178
Betts A (ed) (2011) Global migration governance. Oxford University Press, New York
Biermann F, Siebenhüner B (eds) (2009) Managers of global change: the influence of international environmental bureaucracies. MIT, Massachusetts
Blauberger M, Rittberger B (2015) Orchestrating policy implementation: EU governance through regulatory networks. In: Abbott KW, Genschel P, Snidal D, Zangl B (eds) International organizations as orchestrators. Cambridge University Press, Cambridge, UK, pp 39–64
Bolden R, Hawkins B, Gosling J, Taylor S (2011) Exploring leadership: individual, organizational, and societal perspectives. Oxford University Press, New York
Bradford CI, Linn JF (eds) (2007) Global governance reform: breaking the stalemate. Brookings Institution Press, Washington, DC
Brunsson N (2003) The organization of hypocrisy: talk, decisions and actions in organizations, 2nd edn. Copenhagen Business School Press, Copenhagen
Capoccia G, Kelemen RD (2007) The study of critical junctures: theory, narrative, and counterfactuals in historical institutionalism. World Polit 59:341–369
Chwieroth J (2008) Normative change from within: the International Monetary Fund's approach to capital account liberalization. Int Stud Q 52:129–158
Chwieroth J (2012) "The silent revolution": how the staff exercise informal governance over IMF lending. Rev Int Organ 8:265–290
Chwieroth J (2015) Professional ties that bind: how normative orientations shape IMF conditionality. Rev Int Polit Econ 22:757–787
Cox RW (1969) The executive head: an essay on leadership in international organization. Int Organ 23:205–230
Dijkzeul D, Beigbeder Y (2003) Rethinking international organizations: pathologies and promise. Berghahn Books, New York
DiMaggio PJ, Powell WW (1991) Introduction. In: DiMaggio PJ, Powell WW (eds) The new institutionalism in organizational analysis. The University of Chicago Press, Chicago, pp 1–40
Dingwerth K (2007) The new transnationalism: transnational governance and democratic legitimacy. Palgrave Macmillan, New York
Dingwerth K, Pattberg P (2009) World politics and organizational fields: the case of transnational sustainability governance. Eur J Int Rel 15:707–744
Dingwerth K, Kerwer D, Nölke A (eds) (2009) Die Organisierte Welt: Internationale Beziehungen und Organisationsforschung, 1. Aufl. Internationale Beziehungen 12. Nomos, Baden-Baden
Dingwerth K, Blauberger M, Schneider C (2011) Postnationale Demokratie: Eine Einführung am Beispiel von EU, WTO und UNO. In: Grundwissen Politik, vol 47. Springer VS, Wiesbaden
Dingwerth K, Lehmann I, Reichel E, Weise T, Witt A (2014) Many pipers, many tunes? Die Legitimationskommunikation internationaler Organisationen in komplexen Umwelten. In: Conceição-Heldt E, Koch M, Liese A (eds) Internationale Organisationen: Autonomie, Politisierung, interorganisationale Beziehungen und Wandel. Nomos, Baden-Baden, pp 186–212
Dreiack S (2016) Lernen zwischen Individuum und Organisation: Die Struktur Internationaler Organisationen. Zeitschrift für Internationale Beziehungen 23:105–134
Drezner DW (2007) All politics is global: explaining international regulatory regimes. Princeton University Press, Princeton
Drezner DW (2014) The system worked: how the world stopped another great depression. Oxford University Press, New York

Eckhard S, Mattmüller J (2017) Verwaltungseinfluss und Verhandlungsergebnisse in internationalen Organisationen: Ein ressourcenbasiertes Modell. dms-der moderne staat-Zeitschrift für Public Policy. Recht Manage 10:229–250

Ege J (2016) Verwaltungsautonomie in internationalen Organisationen: Eine deskriptiv-vergleichende Analyse. Springer VS, Wiesbaden

Ege J, Bauer MW (2013) International bureaucracies from a public administration and international relations perspective. In: Reinalda B (ed) Routledge handbook of international organizations. Routledge, New York, pp 135–148

Elsig M (2015) Orchestration on a tight leash: state oversight of the WTO. In: Abbott KW, Genschel P, Snidal D, Zangl B (eds) International organizations as orchestrators. Cambridge University Press, Cambridge, UK, pp 65–87

Enkler J, Grohs S (2017) An den Grenzen der Lernfähigkeit? Lernkurven internationaler Verwaltungen nach der Finanzmarktkrise. dms-der moderne staat-Zeitschrift für Public Policy. Recht Manage 10:251–270

Finkle JL, Crane BB (1976) The World Health Organization and the population issue: organizational values in the United Nations. Popul Dev Rev 2:367–393

Finnemore M (2015) International organizations as teachers of norms: the United Nations Educational, Scientific, and Cultural Organization and science policy. In: Abbott KW (ed) The law and politics of international organizations, Historical, political and normative perspectives, vol II. Edward Elgar, Cheltenham, UK, pp 317–350

Fioretos O (2011) Historical institutionalism in international relations. Int Organ 65:367–399

Graham ER (2014) International organizations as collective agents: fragmentation and the limits of principal control at the World Health Organization. Eur J Int Rel 20:366–390

Gutner T (2005) Explaining the gaps between mandate and performance: agency theory and World Bank environmental reform. Glob Environ Pol 5:10–37. https://doi.org/10.1162/1526380054127727

Gutner T (2010) When "doing good" does not: the IMF and the millennium development goals. In: Avant DD, Finnemore M, Sell SK (eds) Who governs the globe? Cambridge University Press, New York, pp 266–291

Haas EB (1990) When knowledge is power: three models of change in international organizations. University of California Press, Berkeley

Hall N (2015) Money or mandate? Why international organizations engage with the climate change regime. Glob Environ Pol 15:79–97

Hanrieder T (2014a) Gradual change in international organisations: agency theory and historical institutionalism. Politics 34:324–333. https://doi.org/10.1111/1467-9256.12050

Hanrieder T (2014b) Local orders in international organisations: the World Health Organization's global programme on AIDS. J Int Relat Dev 17:220–241

Hanrieder T (2015a) The path-dependent design of international organizations: federalism in the world health organization. Eur J Int Relat 21:215–239. https://doi.org/10.1177/1354066114530011

Hanrieder T (2015b) WHO orchestrates? Coping with competitors in global health. In: Abbott KW, Genschel P, Snidal D, Zangl B (eds) International organizations as orchestrators. Cambridge University Press, Cambridge, UK, pp 191–213

Hanrieder T, Kreuder-Sonnen C (2014) WHO decides on the exception? Securitization and emergency governance in global health. Security Dialogue 45:331–348. https://doi.org/10.1177/0967010614535833

Hopgood S (2006) Keepers of the flame: understanding amnesty international. Cornell University Press, Ithaca, NY

Ivanova M (2009) UNEP as anchor organization for the global environment. In: Biermann F, Siebenhüner B, Schreyögg A (eds) International organizations in global environmental governance. Routledge, New York, pp 151–173

Jetschke A, Theiner P (2016) The diffusion of institutional design among regional organizations. https://papers.ssrn.com/sol3/papers.cfm?abstract_id=2735715. Accessed 13 Mar 2019

Johnson T (2013) Institutional design and bureaucrats' impact on political control. J Polit 75:183–197
Johnstone I (2003) The role of the UN secretary-general: the power of persuasion based on law. Glob Gov 9:441–458
Johnstone I (2007) The secretary-general as norm entrepreneur. In: Chesterman S (ed) Secretary or general? The UN secretary-general in world politics. Cambridge University Press, Cambridge, UK, pp 123–138
Jönsson C (1986) Interorganization theory and international organization. Int Stud Q 30(1):39–57
Kennedy D (2007) Leader, clerk, or policy entrepreneur? The secretary-general in a complex world. In: Chesterman S (ed) Secretary or general? The UN secretary-general in world politics. Cambridge University Press, Cambridge, UK, pp 158–182
Kerler M (2007) Triggering world bank reform: when member states, NGOs and learning get important. In: Bauer MW, Knill C (eds) Management reforms in international organizations, vol 6. Nomos, Baden-Baden, pp 133–148
Kille KJ (ed) (2007) The UN secretary-general and moral authority: ethics & religion in international leadership. Georgetown University Press, Georgetown
Kleine M (2013a) Informal governance in the European Union: how governments make international organizations work. Cornell University Press, Ithaca, NY
Kleine M (2013b) Trading control: national fiefdoms in international organizations. Int Theory 5:321–346
Knill C (2001) The Europeanization of national administrations: patterns of institutional change and persistence. Cambridge University Press, Cambridge, UK
Knill C, Bauer MW (2007) Theorizing management reforms in international organizations. In: Bauer MW, Knill C (eds) Management reforms in international organizations, vol 6. Nomos, Baden-Baden, pp 191–200
Knill C, Enkler J, Schmidt S, Eckhard S, Grohs S (2017a) Administrative styles of international organizations: can we find them, do they matter? In: Bauer MW, Knill C, Eckhard S (eds) International bureaucracy: challenges and lessons for public administration research. Palgrave Macmillan, London, pp 43–72
Knill C, Eckhard S, Bauer MW (2017b) International public administration: Ein neuer Bürokratietyp. dms-der moderne staat-Zeitschrift für Public Policy. Recht Manage 10:173–189
Koremenos B (2005) Leadership and bureaucracy: the folk theorem and real folks. Ration Soc 17:35–79
Koremenos B, Lipson C, Snidal D (2001) The rational design of international institutions. Int Organ 55:761–799
Kühl S (2011) Organisationen: Eine sehr kurze Einführung. VS Verlag für Sozialwissenschaften, Wiesbaden
Liese A, Weinlich S (2006) Die Rolle von Verwaltungsstäben internationaler Organisationen. Lücken, Tücken und Konturen eines (neuen) Forschungsgebiets. In: Bogumil J, Jann W, Nullmeier F (eds) PVS-Sonderheft Politik und Verwaltung. VS Verlag, Wiesbaden
Loescher G, Milner J (2011) UNHCR and the global governance of refugees. In: Betts A (ed) Global migration governance. Oxford University Press, New York, pp 189–209
Loescher G, Betts A, Milner J (2012) The UNHCR: the politics and practice of refugee protection, 2nd edn. Routledge, New York
Luhmann N (1976) Funktionen und Folgen formaler Organisationen. Duncker & Humblot, Berlin
March JG (ed) (1989) Decisions and organizations. Basil Blackwell, Oxford
March JG (1990) Entscheidung und Organisation: Kritische und konstruktive Beiträge, Entwicklungen und Perspektiven. Betriebswirtschaftlicher Verlag Dr. Th. Gabler, Wiesbaden
March JG, Olsen JP (1975) The uncertainty of the past: organizational learning under ambiguity. Eur J Polit Res 3:335–358
March JG, Simon HA (1976) Organisation und Individuum: Menschliches Verhalten in Organisationen. Betriebswirtschaftlicher Verlag Dr. Th. Gabler, Wiesbaden

Martin LL, Simmons BA (2013) International organizations and institutions. In: Carlsnaes W, Risse T, Simmons BA (eds) Handbook of international relations, 2nd edn. Sage, London, pp 326–351

Mayntz R (1963) Soziologie der Organisation. Rowohlts Deutsche Enzyklopädie. Rowohlt Taschenbuch Verlag, Reinbeck near Hamburg

McNamara KR (2010) Constructing authority in the European Union. In: Avant DD, Finnemore M, Sell SK (eds) Who governs the globe? Cambridge University Press, New York, pp 153–182

Mehde V (2007) Creating a missing link? Administrative reforms as a means of improving the legitimacy of international organizations. In: Bauer MW, Knill C (eds) Management reforms in international organizations, vol 6. Nomos, Baden-Baden, pp 163–174

Methmann C, Rothe D, Stephan B (eds) (2013) Interpretive approaches to global climate governance: (de)constructing the greenhouse. Routledge, New York

Meyer JW, Rowan B (1991) Institutionalized organizations: formal structure as myth and ceremony. In: DiMaggio PJ, Powell WW (eds) The new institutionalism in organizational analysis. The University of Chicago Press, Chicago, pp 41–62

Neumann IB, Sending OJ (2013) Governing the global polity: practice, mentality, rationality. The University of Michigan Press, Ann Arbor

Nielson DL, Tierney MJ (2003) Delegation to international organizations: agency theory and world bank environmental reform. Int Organ 57:241–276

Nielson DL, Tierney MJ, Weaver CE (2006) Bridging the rationalist–constructivist divide: re-engineering the culture of the World Bank. J Int Relat Dev 9:107–139. https://doi.org/10.1057/palgrave.jird.1800084

Nye JS (2008) The powers to lead. Oxford University Press, New York

Park S (2009) Socialization, the World Bank Group and global environmental governance. In: Biermann F, Siebenhüner B, Schreyögg A (eds) International organizations in global environmental governance. Routledge, New York, pp 91–110

Pattberg P (2009) Private governance organizations in global environmental politics: exploring their influences. In: Biermann F, Siebenhüner B, Schreyögg A (eds) International organizations in global environmental governance. Routledge, New York, pp 223–243

Prakash A, Potoski M (2010) The international organization for standardization as a global governor: a club theory perspective. In: Avant DD, Finnemore M, Sell SK (eds) Who governs the globe? Cambridge University Press, New York, pp 72–101

Reinalda B, Verbeek B (2004) Patterns of decision making within international organizations. In: Reinalda B, Verbeek B (eds) Decision making within international organizations. Routledge, New York, pp 231–246

Rittberger V (ed) (2001) Global governance and the United Nations system. United Nations University Press, New York

Rittberger V, Zangl B (2003) Internationale Organisationen: Politik und Geschichte, 3rd edn. VS Verlag für Sozialwissenschaften, Wiesbaden

Schechter MG (1987) Leadership in international organizations: systemic, organizational and personality factors. Rev Int Stud 13:197–220

Schimmelfennig F (1997) Rhetorisches Handeln in der internationalen Politik. Zeitschrift für Internationale Beziehungen 4:219–254

Schimmelfennig F (2003) The EU, NATO and the integration of Europe. Cambridge University Press, Cambridge

Schimmelfennig F, Sedelmeier U (eds) (2005) The Europeanization of Central and Eastern Europe. Cornell University Press, Ithaca, NY

Schön-Quinlivan E (2007) Administrative reform in the European Commission: from rhetoric to relegitimization. In: Bauer MW, Knill C (eds) Management reforms in international organizations, vol 6. Nomos, Baden-Baden, pp 25–36

Schroeder MB (2014) Executive leadership in the study of international organization: a framework for analysis. Int Stud Rev 16:339–361. https://doi.org/10.1111/misr.12147

Siebenhüner B (2008) Learning in international organizations in global environmental governance. Glob Environ Pol 8:92–116

Starke P (2013) Qualitative methods for the study of policy diffusion: challenges and available solutions. Policy Stud J 41:561–582

Steffek J (2013) Explaining cooperation between IGOs and NGOs - push factors, pull factors, and the policy cycle. Rev Int Stud 39:993–1013

Steffek J (2015) The output legitimacy of international organizations and the global public interest. Int Theory 7:263–293

Steffek J, Hahn K (eds) (2010) Evaluating transnational NGOs: legitimacy, accountability, representation. Palgrave Macmillan, Basingstoke

Tallberg J, Sommerer T, Squatrito T, Jönsson C (2013) The opening up of international organizations: transnational access in global governance. Cambridge University Press, New York

Taylor FW (1919) Die Grundsätze wissenschaftlicher Betriebsführung. Gesellschaft für sozialwissenschaftliche und ökologische Forschung, München

Timmer V (2009) Agility and resilience: adaptive capacity in Friends of the Earth International and Greenpeace. In: Biermann F, Siebenhüner B, Schreyögg A (eds) International organizations in global environmental governance. Routledge, New York, pp 244–263

Trombetta MJ (2014) Linking climate-induced migration and security within the EU: insights from the securitization debate. Crit Stud Secur 2:131–147. https://doi.org/10.1080/21624887.2014.923699

van der Heijden J (2011) Institutional layering: a review of the use of the concept. Politics 31:9–18

van der Heijden J (2012) Different but equally plausible narratives of policy transformation: a plea for theoretical pluralism. Int Polit Sci Rev 34:57–73

van der Lugt C, Dingwerth K (2015) Governing where focality is low: UNEP and the principles for responsible investment. In: Abbott KW, Genschel P, Snidal D, Zangl B (eds) International organizations as orchestrators. Cambridge University Press, Cambridge, pp 237–261

Veit A, Schlichte K (2009) Internationale Organisationen als gekoppelte Arenen: Wieso scheitern State-Builder? In: Dingwerth K, Kerwer D, Nölke A (eds) Die Organisierte Welt: Internationale Beziehungen und Organisationsforschung, 1st edn. Nomos, Baden-Baden, pp 96–116

Weaver C (2008) Hypocrisy trap: the World Bank and the poverty of reform. Princeton University Press, Princeton

Weaver C, Leiteritz RJ (2005) "Our poverty is a world full of dreams": reforming the world bank. Glob Gov 11:369–388

Weber M (2009) Wirtschaft und Gesellschaft: Die Wirtschaft und die gesellschaftlichen Ordnungen und Mächte. Nachlass. Studienausgabe. J.C.B. Mohr, Tübingen

Weick KE (1985) Der Prozess des Organisierens. Suhrkamp, Frankfurt am Main

Zangl B (2006) Die Internationalisierung der Rechtstaatlichkeit: Streitbeilegung in GATT und WTO. Campus Verlag, Frankfurt am Main

Zürn M (1998) Regieren jenseits des Nationalstaates: Globalisierung und Denationalisierung als Chance. Suhrkamp, Frankfurt am Main

# Chapter 4
# Research Design: Operationalizing Security Theory and Organizational Theory

This chapter builds upon the information on the case and the empirical puzzle (see Chap. 2) and it picks up the theoretical discussions of Chap. 3. Thereby, it constitutes the next step toward answering the overall research question. This chapter outlines the research design of the thesis and it introduces the methods that help to carry out the analysis. As a first step, the chapter operationalizes security theory. Hereby, the focus lies on the different depictions of security from a traditional security perspective and a human security perspective. In a second step, this chapter recapitulates the theoretical strands of organizational theory and derives a list of assumptions from the theoretical groundwork of Chap. 3. Following these preliminary steps, the chapter presents the processes of data collection and data analysis. Therefore, this chapter elaborates on why this thesis uses the methods of qualitative expert interviews and qualitative content analysis.

## 4.1 Operationalizing Security Theory

This section builds upon the approaches of security and organizational theory introduced in Chaps. 2 and 3 and operationalizes these different theoretical 'lenses'. For this thesis, the theoretical assumptions derived from security research mainly focus on how the UNHCR refers to EDPs. Consequently, the focus lies on what kind of depiction the UNHCR uses. Therefore, this strand of research is not split into various theoretical assumptions or lists of questions. Instead, the analysis centers on whether the UNHCR ideationally depicts EDPs in reference to traditional security or in reference to human security. These contrasting connotations of security give a clue on how the UNHCR intends to deal with EDPs.

First, if the organization's documents and speeches refer to EDPs in terms of a danger to societies and national or international security, the timeframe will be classified as traditional security. The wording of the UNHCR might hereby depict EDPs as a threat to the nation state or to the safety of its population. Also, those

 
S. Hantscher, *The UNHCR and Disaster Displacement in the 21st Century*, Contributions to Political Science, https://doi.org/10.1007/978-3-030-19689-9_4

people who are affected by environmental changes might be presented in an excessive and disproportionally high number. Overall, the psychological effect of this kind of security reference on the receiving countries is huge and there is a high chance that the receiving population develops a defensive demeanor toward EDPs. For this thesis, the traditional interpretation of security relies on how the term was originally understood by the Copenhagen School or Paris School.

Second, if the organization's documents and speeches argue in terms of individual catastrophes and restrictions to human livelihood, this will be categorized as human security. As this depiction of EDPs puts the human being central, the psychological effect on the receiving population is a different one when compared to the traditional conception of security. By referring to EDPs as individuals who suffer due to environmental catastrophes and who must leave their homes because of unbearable conditions, the people will likely be perceived as helpless victims. If it becomes apparent that the UNHCR refers to this kind of security, the message it sends out to the world is a very different one. Being a humanitarian organization itself, the wording surrounding human security comes naturally to its own organizational agenda and will thus not be viewed as extraordinary.

The term 'ideational' hereby refers to Constructivism because the term is based on the benchmark works of Kratochwil and Ruggie (1986, p. 774) who were the forerunners in proposing more interpretative approaches when it comes to international regimes and IOs. The 'ideational' aspects in this theoretical approach thus depend on the ability of IOs to "enhance (or diminish) intersubjective expectations and normatively stabilized meanings" (Kratochwil and Ruggie 1986, p. 775).

Table 4.1 depicts the strands of security theory that have been introduced above (see Sect. 2.1) and the table operationalizes these theoretical perspectives for the purpose of this thesis. Hereby, the focus lies on those aspects that need to be recognizable in the UNHCR's security references before one can speak of a traditional security depiction or a human security depiction. In addition to the security references of the UNHCR, the focus also lies on the organizational factors that might explain the empirical puzzle. To tackle these external as well as internal factors, the following section shows the operationalization of organizational theory.

**Table 4.1** How to recognize the ideational depiction of traditional security or human security

| Operationalization of (traditional) security | Operationalization of human security |
|---|---|
| • Focus on the nation state or the international community as a whole | • Focus on the individual |
| • Threat to livelihood of the receiving population | • Threat to livelihood of the affected people |
| • Details on the victims: their poverty, wickedness, and huge number | • Details on the victims: their suffering and helplessness |
| • Solutions: traditional security connotations, e.g. the military | • Solutions: plea for humanity, development aid, admission of refugees |

## 4.2 Operationalizing Organizational Theory

This section outlines the relevant theoretical concepts derived from organizational theory. Hence, the chapter presents several theoretical assumptions. All of these assumptions have a say and correlate with the overall research question, namely why the UNHCR approaches the topic of EDPs the way that it does. These assumptions represent individual research questions in their own right and will thereby help to build the interview questions. The input from organizational theory is hereby subdivided into a meaningful manner with regards to content. Starting with the external factors such as, first, the organizational environment and, second, the member states, the focus shifts toward more intra-organizational facets such as, third, the character of the bureaucracy and, fourth, the intra-organizational strategies.

### *4.2.1 The Organizational Environment*

The first strand of theoretical assumptions from organizational theory is derived from research that focuses on the environment in which an IO acts. As discussed in Sect. 3.3.1, arguments concerning global governance, the relationship with other IOs and external factors fall under this theoretical frame. The central assumption that is derived from this strand of research is that the organizational environment influences IOs and that an IO's actions cannot be interpreted in isolation. Rather, organizational research expects organizations to act in response to external push factors that explain why IOs change and why they take up new issues. In relation to the empirical case study of this thesis, external factors are thus expected to be the source of UNHCR's interest in EDPs.

That said, the research on the organizational environment expects the UNHCR's interest in EDPs to result from inter-organizational networks. Concurrent to the state of the art outlined above, IOs form coalitions on certain matters of interest, or they compete over resources and influence. The research on the organizational environment thus states that the IOs that also work on human mobility might either be considered as push-factors for the collaboration or the UNHCR might consider them to be rivals. For this reason, the state of the research predicts that the IOM, the NRC, and the IDMC play a role in explaining why the UNHCR approaches the issue of EDPs the way that it does.

Additionally, the research on the organizational environment mentioned in Sect. 3.3.1 also makes statements about the difference between supranational and international governmental organizations. Whereas supranational organizations lack national interference, international governmental organizations largely rely on member state preferences, which might lead to a stronger and much quicker impetus for change. This strand of research thus forecasts that the broader organizational environment—e.g. national perceptions of the issue of EDPs or a global willingness to tackle the issue of environmental migration—influences UNHCR decision-making.

Consequently, the following list of assumptions can be derived from this field of research:

1. The UNHCR's organizational environment influences its approach toward EDPs.
   (Does the UNHCR's organizational environment spark its approach toward EDPs?)
2. Inter-organizational relations influence the UNHCR's approach toward EDPs.
   (Do inter-organizational relationships push the UNHCR's EDP approach?)

### *4.2.2 The Member States*

The second strand of theoretical assumptions is derived from research that focuses on the role of member states in an IO's decision-making. As outlined in Sect. 3.3.2, the role of member states is often underestimated, and the behavior of IOs is interpreted as simply a reflection of the organizational agenda. Nevertheless—and following the current state of organizational research—it is quite common for states to play a strong and influential role in the daily work of IOs. This becomes apparent in IOs' administration and in the budgetary structure, which might mirror member states' wishes and control. In relation to the empirical case study of this thesis, member states might be the source of the UNHCR's interest in EDPs.

The assumptions that can be deduced from this strand of organizational research build upon the principal-agent approach. Arguing from this line of thought thus leads to the assumption that the role of states in an organization's decision-making is pivotal and that an IO's agenda is soaked with national interests. This reading of IOs thus allows national power struggles to be taken into consideration. Additionally, if an IO strikes a new path, the principal-agent research will attribute this to member states' interests. For the case study of the UNHCR and its agenda toward EDPs, the principal-agent approach assumes that the relatively sudden interest in matters of environmental migration arise from the member states. As this theoretical lens sees IOs as agents that are governed by national interests, the interest in EDPs is expected to originate in the UNHCR's member states, with or without only a small contribution coming from the organization itself. Furthermore, this strand of research also assumes that the member states preside over all budgetary matters of an IO. Consequently, states play the role of policy brokers, and the administrative body of an IO depends on what the member states decide. For the case study at hand, this implies that the member states play a powerful role in the UNHCR's organizational structure, and thus, with respect to its finances, the UNHCR's approach toward EDPs has been spearheaded by member states and national interests.

Lastly, the current research on the role of member states expects that there are national fiefdoms in the internal structure of IOs. This implies that certain departments or sections are implicitly or explicitly controlled by member states or coalitions of member states. In the case of the UNHCR, the interest in EDPs might thus be ascribed to national interests that are enforced through certain UNHCR bodies or committees. As such, the principal-agent mindset still applies, because the principals influence the agent just the same, although this does not happen in a strict top-down manner, but rather from within the organization. The organizational structure of the UNHCR might thus show spheres of member states exercising influence that explain certain decisions that do not flow naturally from the UNHCR's mandate. Consequently, the following assumption can be derived from this field of research:

3. The UNHCR's 'member states' have an influence on the organization's EDP approach.
   (Do states control and push the UNHCR's EDP approach?)

### 4.2.3 *The Character of the Bureaucracy*

The third strand of theoretical assumptions is derived from the research that focuses on the organizational bureaucracy. The condensed reproduction of the current state of research on the character of the bureaucracy conveys that these factors are important when confronted with a puzzling organizational behavior. In relation to this thesis, the organization's bureaucracy is expected to be the source of the UNHCR's interest in EDPs.

As mentioned above, the opening up of the 'black box'—the internal DNA—is a response to the principal-agent approach, which denies IOs any kind of actorness. When analyzing the UNHCR's policy decisions, the choices made on the matter of EDPs are thus ascribed to the organization itself. The research done on the administration of IOs therefore expects that IOs act in an atmosphere where there is a scarcity of time and a multitude of short- and long-term goals. This strand of research expects that the decision to consider EDPs might need to be interpreted against certain short- or long-term goals of the UNHCR's administration and should thus not be construed in isolation from the UNHCR's overall goals.

Furthermore, this branch of research anticipates that IOs act as agenda-setters and that the decision to push a new issue internally or externally might reflect the short-term goal of achieving this role of agenda-setter. This theoretical lens on organizational research stresses that the interest of the UNHCR in EDPs might originate from an urge to take on the role of an agenda-setter on the issue—and that the organization, thus, is not necessarily aiming to include the protection of EDPs in its long-term goals. From this theoretical perspective, the UNHCR becomes an actor on its own that has authority over and leads member states' policies toward the topic of environmental migration.

The research on the administration also expects inter-organizational differences of opinion. That said, the headquarters of the UNHCR in Geneva might take another stance on the issue of EDPs than the field offices. From the perspective of administration research, the field offices might be the arena inside the organization that triggered the interest in EDPs. Then again, the issue of EDPs might have been pushed from the headquarters, without considering the field offices' realities on the ground. In any case, this theoretical strand of organizational research expects a difference in opinion inside the UNHCR and distances itself from research that considers the UNHCR as a homogenous actor. Hereby, the focus also lies on the workforce inside IOs. In the case of the UNHCR, this implies that certain individuals inside the UNHCR might have a leading role in pushing the issue of EDPs for numerous reasons, such as their own careers or their reputation. The question arises of how the UNHCR's funding is organized, as this provides hints as to what role the member states play and what kinds of mechanisms of influence individuals inside the UNHCR have at their disposal.

Organizational research on the bureaucracy also assumes that the executive head of an IO might play an important role when it comes to striking a new path. Given that executive heads aim to satisfy the member states' wishes, this strand of research expects that they will only make small and pre-determined organizational changes. Organizational research thus assumes that the decision to speak out on EDPs is based on the lowest common denominator, that is, it is accepted by most of the participants in the UNHCR. Also, the executive head—in the case of the UNHCR, the High Commissioner—is highly dependent on the expertise of the specialized departments and sections. As stated by the research on executive heads, the work of the staff might steer the superior in a certain direction. The hierarchical structures of the UNHCR are thus helpful in interpreting the decisions made on the matter of EDPs. For the case at hand, the current state of research implies that the High Commissioner approves of the UNHCR's approach toward EDPs. In other words, it is unlikely that the matter of EDPs emanates from within the organization without the previous consent of the High Commissioner. Consequently, the following assumptions can be derived from this field of research:

4. The UNHCR approaches the topic because it overlaps with the organization's goals.
   (Does the issue of EDPs relate to any of the UNHCR's goals?)
5. The UNHCR's headquarters push the organization's EDP approach.
   (What roles do the headquarters and the field offices assume in relation to the topic?)
6. Individuals in the organization push the UNHCR's EDP approach.

(continued)

(Do individuals inside the UNHCR take a leading role on the matter of EDPs?)

7. The High Commissioner pushes the approach and has a personal motive for doing so.

   (What position and role does the High Commissioner assume on the question of EDPs?)

### 4.2.4 The Intra-Organizational Strategies

The fourth strand of theoretical assumptions is derived from the research that focuses on intra-organizational strategies. One assumption is that IOs are aware of their public perception, and thus, aim to put forth a certain image of themselves to the outer world. When interpreting the UNHCR's approach toward EDPs, it thus becomes paramount to identify the desired image that the organization wants to convey to the broader public. This theoretical strand thus shifts the focus toward the hidden realities behind the corporate design of the UNHCR, and it tries to uncover the discrepancies between the public image and the organization's daily business. By means of window dressing the organization, the UNHCR is able to protect its hidden agendas and its long-term plans, both of which the analysis at hand wants to elucidate. The intention is to uncover whether the UNHCR is also showing signs of this kind of organizational behavior, and if it is, what can be drawn from this. This theoretical strand of research thus expects that the UNHCR uses window dressing to convey a certain image to the outside world.

Concerning the role of individuals, this theoretical lens assumes that the individuals inside IOs learn from previous experiences and that they consider these learning processes during decision-making. This strand of research thus expects that the interest of the UNHCR in EDPs is the outcome of a learning process conducted inside the organization. It is thus essential to identify previous decisions that might have cleared the path for the UNHCR's EDP approach. The internal organizational rules and procedures influence the individuals who work at the UNHCR, and these aspects guide them in their decision-making process. This theoretical strand expects that there exist previous developments that triggered the UNHCR staff to rethink their decisions. Therefore, 'learning' of the UNHCR workforce implies that while discussing EDPs, the staff recaps earlier decisions—i.e., the staff considers previous decisions in the context of EDPs. In order to be able to speak of a learning process by staff, it would need active implementation of the topic in the respective divisions. A simple briefing of the media personnel and the regional offices does not fulfill the criteria for a learning process.

The last intra-organizational strategy is summarized as organizational change. This theoretical 'lens' expects that the individuals working at the UNHCR strive for change and renewal. Also, it is assumed that a final trigger is needed—usually in the form of external pressure—for the change to become institutionalized. Leaning on the assumptions deduced from organizational theory, organizational change is ascribed to dissatisfied member states that intend to push the organization in a new direction (rationalist institutionalist). For the case study at hand, this theoretical 'lens' thus assumes that member states are behind the UNHCR's interest in EDPs. Additionally, the theoretical lens of organizational change expects that the interest in EDPs must be understood in terms of an incremental organizational change over time (historical institutionalism). Therefore, the approach of the UNHCR was not triggered by sudden events, but rather mirrors an overall tendency of the organization to change. This organizational theory assumes that the interest in matters of EDPs can be attributed to long-ranging changes inside the organization's bureaucracy. From this perspective, the founding moment of the UNHCR becomes important, as this period of the organization might predict the interest in new issue matters—e.g. EDPs—later. As the establishment of the UNHCR was a state-led process, the member states are very likely to have been influential throughout the overall existence of the UNHCR.

Furthermore, this organizational 'lens' expects critical junctures in the life span of an organization. This implies that before the UNHCR's interest in EDPs, there must have been incidents that paved the way to a relaxation of the normal organizational structures. It is thus necessary to reconstruct possible path-dependent processes for which the matter of EDPs depicts the peak. In the context of this thesis, 'organizational change' implies that there have been incidents that paved the way to the UNHCR's interest in EDPs and that one can identify long-ranging and distinct changes inside the organization's bureaucracy due to these initial incidents. Consequently, the following list of assumptions can be derived from this field of research:

8. The EDP approach is a means to enhance the UNHCR's public image.
   (What kind of public image does the UNHCR want to convey? Are there hidden realities behind the UNHCR's façade?)
9. The UNHCR's EDP approach represents a long-term learning process.
   (Does the EDP approach reflect an organizational learning process?)
10. The UNHCR's EDP approach depicts a change in the organization.
   (Does the UNHCR's EDP approach signify an organizational change?)
11. The UNHCR approaches the issue because it correlates with its norms and values.
   (Does the interest in EDPs emerge from the UNHCR's normative foundation?)

## 4.3 The Processes of Data Collection and Data Analysis

The aim of the research project is to uncover the hidden motives and incentives of the UNHCR to approach the topic of EDPs the way that it does. As it is hard to uncover the internal workings without a glimpse inside the organization, the 'black box'—the organization itself—needs to be opened with the help of insider perspectives as well as other IOs' perspectives on the UNHCR. To fulfill the scientific standards, it is necessary to lay open the process of data collection, to explain the choice of experts, to reveal the set questions, to outline the interview situation, and to present the process of data analysis.

### *4.3.1 Qualitative Expert Interviews*

The following paragraphs present the whole process, from choosing and contacting the experts to structuring the interview into set questions and, finally, conducting the expert interviews in Geneva, Switzerland. Overall, the method of qualitative expert interviews falls into the category of qualitative research. This implies that the method is not sufficient in coming to overall conclusions or making generalizations. Instead, the method of qualitative expert interviews is chosen, because it helps to understand the empirical puzzle (the UNHCR's EDP approach) based on a systematic analysis (Kaiser 2014, p. 71). Therefore, the expert interviews do not need to fulfill the criterion of a representative sample.

In expert interviews, the level of standardization is relatively low, and even though an expert interview has a half- or even fully structured format, it is difficult to evaluate statistically (Kaiser 2014, p. 3). The (at least) half-structured format is necessary to extract information by means of a series of set questions. Consequently, to analyze an expert interview means to use interpretative methods of data analysis. Expert interviews are in-depth studies and aim to show the specific characteristics of the single case. In other words, expert interviews help to understand a case study, as opposed to quantitative methods, which try to generalize (Kaiser 2014, p. 4). For a scholar who uses expert interviews in his or her research, the case study is the central object of analysis, and he or she either describes the case at hand or reconstructs mechanisms that might explain change in the case study.[1]

[1]Qualitative expert interviews that are used in case study analyses can oppose quantitative analyses that use a more standardized means of analysis. Whereas the latter do not need a face-to-face interaction with the chosen expert, the first depends on direct contact between the expert and the interviewer (Kaiser 2014, p. 1). In case study analyses, expert interviews aim to uncover concrete information. The format of the expert interview thus differs from ethnographic interviews utilized in participatory observations or narrative interviews that focus on historic case studies and biographies (Kaiser 2014, p. 3). Rather, political scientists use theory-informed expert interviews to analyze political behavior or to shed light on the decision-making behind policy decisions (Blatter et al. 2007, p. 62).

For this thesis, an expert is defined as someone who has special knowledge about a social issue, and a qualitative expert interview is *the* method to uncover this knowledge (cf. Gläser and Laudel 2004, p. 10). Accordingly, an expert is someone who, first, has a certain position and specific knowledge surrounding this position and, second, someone who has expertise on the concrete information that is relevant for the scholarly analysis (Kaiser 2014, p. 38). The experts themselves are thereby not the primary objects of scholarly interest; instead, the experts are witnesses of the processes in which the researcher is interested (Gläser and Laudel 2004, p. 10). To put it differently, this thesis uses the expert interviews to reconstruct organizational decision-making; the interviews make the organizational decision-making accessible to the researcher.

The approach of a research project that uses expert interviews as a method of analysis is as follows: In a first step, the researcher considers theories—i.e., theoretically based explanations for the puzzle at hand—and applies the theoretical mechanisms to the chosen case study. Then, the researcher fulfills a second step in which he or she operationalizes the theoretical questions into an empirically fitting set of questions. The interviewer confronts the interviewee with operationalized questions that are of concern to the interviewee's work, experiences, etc. In a third step, the researcher reapplies the information gained from the expert interviews to the theoretical context of the analysis (Kaiser 2014, pp. 4–5).

#### 4.3.1.1 Choosing the Experts

The qualitative expert interview method does not imply that all relevant experts must be interviewed (Kaiser 2014, p. 71). Instead, the interviewer chooses the interviewees for an expert interview based on contextual consideration. The choice and number of experts depends, first, on the specific research question and, second, on the research design of the project.[2] Expert interviews are an appropriate methodological choice when it comes to single case studies that need to be analyzed in a detailed manner and from various perspectives (cf. Blatter et al. 2007, p. 60). In the field of environmental migration and displacement, there are only a few IOs that take a center position and that are key influencers. These IOs are the IOM, the NRC, the Nansen Initiative, the IDMC and the UNHCR. Hereby, every IO has its specific area of expertise, follows its agenda and captures a position in the field of environmental migration. However, working at the interphase of the highly competitive topic of EDPs makes it difficult for an IO to act on its own. As a result, the IOs take each other's work into consideration and they largely cooperate and work together.[3]

[2]Hereby, the decision on the number of expert interviews is highly context-specific, and in the academic literature, there exist no general rules about the sufficient number of expert interviews (Gläser and Laudel 2004, p. 102).

[3]However, the field of environmental migration and displacement is contested, and as such, it is to be expected that every IO must fight for funding and its external acceptance. Thus, it would be too

The chosen experts for the qualitative expert interviews are therefore representatives of the IOs mentioned above. In addition to the organizational staff, the thesis also incorporates interviews with scholars from the field of environmental migration and displacement. The aim is to get a picture of how fellow academics who have worked together with the UNHCR on the topic of environmental migration interpret the behavior of the organization. The academic experts have a long working relationship with the UNHCR and thereby dispose of the expertise to talk about the intra-organizational processes. To complement the row of experts, the voices of state representatives are included. Here, the focus is, first, on developed states, which are frontrunners on the topic of EDPs and which try to push the topic in international forums. Second, the focus is on developing states that suffer from the impacts of environmental changes and thus have a strong impetus to raise awareness of the topic. The analysis includes voices from both types of states by means of a number of expert interviews with state representatives in Geneva.

Due to the relatively recent occurrence of the topic of EDPs in the international sphere, the number of people who have been involved with the topic inside the IOs is manageable.[4] This thesis defines an 'expert' as someone who possesses certain knowledge that exceeds common knowledge, and this expert knowledge emanates from his or her function and role in the IO (Kaiser 2014, p. 36). As some of the questions refer to the respective influence and own career desires of the international civil servants, special care has to be given to the validity of interview data (as discussed by Biermann et al. 2009, p. 63, who follow a similar path of research). Consequently, it is to be expected that the experts might talk about their own work very positively and that they might ascribe themselves a lot of influence on the matter. Therefore, the set of questions for the expert interviews intends to uncover the actual workings in the organization and hence tries to coax the truth out of the experts.

Building upon this scholarly discussion on the chosen method, the focus now shifts back to the UNHCR. To begin, the organizational structure of the UNHCR gives a first idea of which departments in the organization deal with the issue at hand. As I have already met UNHCR staff who work on the matter of environmental migration and displacement at international conferences, I could directly contact them via email by referring to the previous meeting. These first interviews thus led to more interviews at the UNHCR headquarters in Geneva, Switzerland, because the experts recommended other people who have worked on the topic at the UNHCR.[5]

early to speak of a fully elaborated collaborative network between these international actors. Rather, the study at hand will shed light on how the IOs interact and to what extent and why they cooperate.

[4]The choice of experts also illustrates one source of error: the researcher only has limited influence on the fact that the expert might—knowingly or unknowingly—give wrong information. After the interviews, the scholar needs to differentiate between official statements that can be found in organizational publications and between unofficial, new and project-relevant information (cf. Kaiser 2014, p. 33).

[5]The first interviews have been helpful in a twofold way: first, the experts at the UNHCR recommended other colleagues at the UNHCR; second, the experts could outline the organizational

Still, it became obvious that the UNHCR does not have a very long history of dealing with the topic; thus, the number of people who worked on the topic at the UNHCR is relatively small. This insight is important for the interview process and gives a first clue about how the UNHCR approaches the topic of EDPs the way that it does.

The other experts were contacted via email. All emails were written in the same style: in the first paragraph, I shortly introduced myself and I referred to the first interviewees as a reason why the person received this email; in the second paragraph, I shortly described my research project and already connected the respective person's professional background to the project's content; in the third paragraph, I guaranteed the potential interviewee anonymity (cf. Kaiser 2014, p. 78).

#### 4.3.1.2 Guided Interview

One central aspect of the guided interview is how the interviewer interrogates and questions the interviewee. The set of questions needs to allow for all kinds of answers and should avoid leading the interviewee into one direction. The interviewer must be open toward new findings that might contradict what he or she expected prior to the expert interview. That said, the set of questions also intends to uncover the individual background of the interviewee: The career path and moral concepts of the expert are important background information that help to interpret the interview data (Kaiser 2014, pp. 8–9). In this context, the interviewer also tries to uncover whether the expert only gives answers that he or she assumes to be socially acceptable (Kaiser 2014, p. 33). These challenges reflect the ongoing discussion about expert interviews and show why there exists no standardized approach or generally acceptable procedure for qualitative expert interviews.

The chosen set of questions for this analysis is intended to give the interviewee the chance to tell his or her exclusive expert knowledge. The researcher should therefore *not* ask the interviewee the research question up front or explain the empirical puzzle of the research project (Kaiser 2014, p. 41). Instead, the theoretically derived assumptions are transferred into interview questions for the chosen experts and each interview question refers to one possible theoretical explanation for the empirical puzzle. The set of questions structures the interview situation and provides a benchmark and a chronological order (cf. Kaiser 2014, p. 52; Westle 2009, p. 251) and the theoretical background helps to develop the interview questions. Consequently, the interviewer knows which block of questions in the interview refers to which kind of theoretical assumption. Following the interviews and during the analysis of the data, the interviewer needs to extract the relevant information. Despite these structural reasons, the order of the guided interview also intends to show the expert that the interviewer has expertise on the topic. As

structure of the UNHCR and give more background on the staff structure. This second aspect is relevant, because the organigram of the UNHCR only tells half the truth about the internal workings of the organization.

shown in the set of questions, the questions that have a rather general direction are at the beginning of the interview, and the more detailed questions follow in the middle of the interview.

Following the first contact via email and the concurrent communication concerning the date for the interview, the actual interviews took place in Geneva, Switzerland in spring 2016. At the beginning of the interview, the purpose of the expert interview was outlined and the interviewee was introduced to the research project. Additionally, the interviewee was informed that the interviews would only be cited anonymously in the publication.

#### 4.3.1.3 Set of Questions

The set of questions for the interview are chosen and asked in a predetermined manner (see Fig. 4.1). The aim is to have an open, trustworthy and gainful interview situation for the interviewer and the interviewee.[6] The interview starts with a short but self-confident introduction of the interviewer and the research project at hand. The interviewee is informed that all the information given in the interview will be handled carefully and anonymously.[7] Additionally, the interviewer asks if she can use a voice recorder. In these first few minutes of the interview, the interviewer outlines the structure of the interview. The guided interview includes different types of questions[8]: introductory questions, context questions, background questions and questions about the opinion of the expert (Gläser and Laudel 2004, pp. 118–126). During the guided interview, these occur in a mixed way, depending on the intention underlying the question and depending on the interview situation at hand. Nevertheless, the structure of the interview questions follows the contextual circumstances; that is, those questions that belong to one context will also be asked together (Gläser and Laudel 2004, pp. 140–142).

---

[6] Building on previous work on qualitative expert interviews, the interviewer is aware of possible hindering effects during an interview (cf. Kaiser 2014, p. 81). It is important to avoid an interview situation in which the interviewer and his or her project are perceived as untrustworthy or in which the interviewer shows disinterest in the interviewee. If the interviewee dominates the interviewer and tries to communicate those matters that he or she perceives relevant, the outcome of the interview will be unsatisfying (Kaiser 2014, p. 81). Similarly, the expert might use the situation for his or her self-portrayal to push his or her self-confidence. Also, it is important to keep in mind that the daily business of the expert, the way he or she interprets the IO's decisions, is very different to the scientific context through which the interviewer sees the situation (Gläser and Laudel 2004, p. 108).

[7] All these background information and organizational details intend to show professionalism and hopefully give the interviewer an accepted position in the following communication.

[8] The set of questions of the expert interviews are problem centered. A problem-centered interview builds upon a theoretical background, and this type of interview is aimed at discussing previously made theoretical assumptions (Blatter et al. 2007, pp. 61–62). The expert interview that is conducted with a set of questions is one example of a problem-centered interview.

**Interview date: __________; expert:____________________________; IO:_________**

**Info on the dissertation project and on the interview:**

Anonymous; voice recorder

First block of questions is on you and your career at the UNHCR.

Second block is on your work and on EDPs.

Third block is on the organization itself—on the UNHCR.

**Block A:**

1. For how long have you been working at the UNHCR?
2. Where did you work before?
3. Have you already worked at different divisions/departments at the UNHCR? Positions?
4. In which division do you work right now?
5. What were the most important changes during your work at the UNHCR?
6. Which global challenges influence the work of the UNHCR most? And in the future?
7. Are there any specific moral concepts (ideals) that dictate the daily routine?

**Block B:**

8. During your work at the UNHCR, did you deal with natural disasters and climate change as causes for people leaving their homes?
9. When did you first come across the two topics in your daily work?
10. When did the UNHCR first approach the issue of disasters and climate change? Did the topic stay on the (UNHCR) agenda?
11. How does the UNHCR approach the issue of EDPs?
12. Who started it? Was there some kind of coalition behind it? Which role did states play?
13. What do you think are the central points of the agenda?
14. Was the topic controversial? Who was against it? And why? (Clashes of funding?)
15. Why does UNHCR approach the issue at all?
16. Is it top priority? And in the past? Has it been different?
17. How do you explain this shift in priority?
18. Who talks about the topic at UNHCR? Who is 'the voice' on the topic?
19. Which other departments speak out on the topic? Do you cooperate with them?
20. Which role do the field offices play? Is there some kind of feedback loop?

**Fig. 4.1** Set of questions for the qualitative expert interviews

**Block C:**

21. What is your personal experience with the cooperation with other IOs?
22. Implementation of new policies: Which role does your department play and which role do states play?
23. In your daily work, which role do states play?
24. If you want to start a new project or you want to change something, what do you think are your chances of doing so?
25. How do you cope with conflicts inside the UNHCR?
26. Concerning career paths: Is the UNHCR staff recruited from other UN-organizations and changes frequently back and forth between IOs, or do most people remain at the UNHCR?

**Block D:**

Anything else you want to add or you want to talk about?

Can you recommend some colleagues?

**Fig. 4.1** (continued)

Building upon this introductory phase of the interview, the questions in block A of the set of questions in Fig. 4.1 are the warm-up questions. These questions are relevant for the overall research question, because they give the interviewer the chance to take the personal background and career path of the interviewee into consideration. The last three questions of block A (5, 6 and 7) already constitute the transition to block B: question 5 and question 6 ask about the organization for which the interviewee works and hint at possible changes inside the IO. Question 6 specifically refers to possible organizational challenges with which the staff is confronted and also gives the interviewee the opportunity to explain what kind of challenges he or she expects for the organization in the future.

Blocks B and C of the set of questions focus on the topic of the research project, namely EDPs and the UNHCR's approach toward the topic. These blocks consist of direct questions, indirect questions and questions that structure the interview. The aim of block B is thus to get the expert talking on the topic of EDPs from his or her respective organizational background. Throughout the discussion of the questions of blocks B and C, the interviewer steers the attention to the UNHCR and tries to uncover the hidden motives of the UNHCR. In the expert interviews with the UNHCR staff, the questions directly ask about the interest of the organization in the topic: When did the topic first occur inside the organization? How does the expert summarize the UNHCR's approach on the topic? Who is—according to the expert—the spokesperson on the topic inside the UNHCR? Questions 21–26 specifically tackle the different theoretical assumptions that I have developed before the interview and which I deduced from organizational theory. The middle part of the interview is therefore characterized by more in-depth questions that each have one theoretical perspective or argument in mind. Behind every interview question is a specific theoretical background, but this is hidden from the interviewee.

During the last part of the interview, in block D, the interviewer gives the interviewee the chance to add to what has already been said and to recommend other experts. Generally speaking, in guided interviews, the interviewer is relatively free in how to formulate the interview questions during the interview situation, and the researcher is able to handle the order of the questions rather flexibly (Gläser and Laudel 2004, p. 39). Depending on the specific interview situation, it might also be necessary to ask an additional ad hoc question to get a concrete answer for the original question.

#### 4.3.1.4 Voice Recording and Transcription

All expert interviews are recorded with a voice recorder, which constitutes the most common procedure for qualitative expert interviews (cf. Kaiser 2014, p. 84). The disadvantage of a voice recorder is that the interviewee might feel intimidated and might thus restrict him- or herself in speaking openly. Especially those interviewees who are not familiar with an interview situation might be afraid that the content could leak to their superiors or colleagues. Still, it has been shown that the interviewees usually forget about the recorder throughout the interview, which relativizes the above-mentioned negative effects (Gläser and Laudel 2004, p. 153). The advantages of a voice recorder outweigh the disadvantages for this specific research project: as I worked on my own, it would have been unsatisfying to make only some written notes throughout the interview—especially because these notes only give a hint of an idea of the interview content.[9] Adding to the interview transcripts, the interviewer also hand-wrote a protocol after each interview (cf. Gläser and Laudel 2004, p. 187; Kaiser 2014, p. 88). These protocols often included a short note on the biographical background of the interviewee. Also, I wrote down my first impressions about the interview situation and about the atmosphere throughout the interview. It is also important whether the expert showed more interest in some of the topics as compared to others and whether he or she gave the names of fellow colleagues to contact for future interviews.

Due to reasons of anonymity, the names of the experts are not mentioned in the analysis. Nevertheless, the scientific criterion of intersubjective verifiability must be fulfilled throughout the research (cf. Kaiser 2014, p. 86). Consequently, the analysis at least shows whether the information is gained through an interview with a representative of an IO or NGO (‘IOR’ or ‘NGOR’), a state representative (‘SR’), or with an academic (‘AR’). The anonymity of the experts is an ethical basis of the

[9]This disadvantage only declines if two interviewers conduct the expert interviews; one interviewer can communicate and focus on asking the questions, and the second interviewer can write notes and focus on how the questions are answered.

research project at hand, because the interview questions directly relate to the inner workings of the organization and focus on the role of certain individuals inside the IO. It is a high priority to guarantee that the experts keep their privacy and that their participation in the interviews does not impact their future careers (cf. Gläser and Laudel 2004, pp. 48–53). All interviews were fully transcribed with the help of the qualitative data analysis software MAXQDA.

### 4.3.2 *Qualitative Content Analysis*

The qualitative content analysis of the expert interviews uses categories that help to approach the texts. These categories arise from the theoretical concept of this thesis. The analysis also presents one new category that arose throughout the data collection phase. The literature review of organizational theory did not anticipate this category, i.e., this possible explanation for the UNHCR's EDP approach. Supposing that a certain argument or hypothesis shows in more than one of the expert interview transcripts, they function as a new category for the analysis and thus are incorporated into the coding system. Because of this approach, the coding and categorizing occurs in a deductive as well as inductive manner, which is an approach that other political scientists have already applied (cf. Powell et al. 2012, p. 244). Generally speaking, the qualitative content analysis of the expert interviews occurs in the following three steps (see Fig. 4.2; building upon Kaiser 2014, pp. 89–110).

First, the transcribed interviews are uploaded to the software-program MAXQDA, and the interviewer bluntly codes the material; that is, it is checked whether the interview holds paragraphs with important information for answering the central research question. In other words, the material is skimmed and reduced for the next coding step.

Second, the interviewer considers the reduced text material from step one and identifies the core message of the respective paragraph or text.[10] This second step takes the theoretical assumptions—categories—into consideration, which the interviewer reduced from theory and which were already used as a guideline for the set of questions (Blatter et al. 2007, p. 178). Therefore, the second step in the data analysis of the expert interviews is rather deductive. The problem of reliability of data is not a big problem in the context of this analysis: Just one person interviewed all experts, which means that the problem of different ratings in the categorizing and coding process is minimal (cf. Biermann et al. 2009, p. 64). As this second step of the content analysis explicitly uses the theoretical concept that the interviewer built prior to the interviews, this step can be referred to as theoretical coding (cf. Blatter et al. 2007, p. 75).

[10]The reduction of the interview material is one criteria of a qualitative content analysis as opposed to a discourse analysis. The latter does not reduce the interview material but rather tries to find an overall motive and meaning of the discourse at hand (Blatter et al. 2007, p. 76).

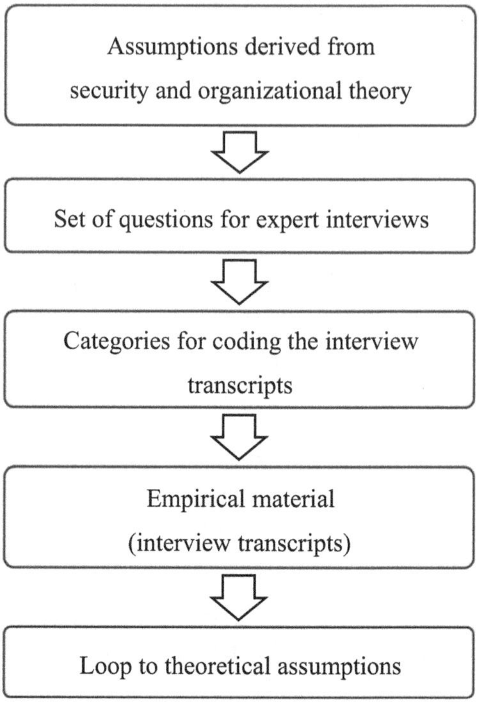


**Fig. 4.2** Model of the data analysis for the qualitative expert interview transcripts

Third, the core messages of all interviews are marked in MAXQDA and put into relation with each other. During this step, it is important to differentiate between the answers of the experts and to acknowledge to what extent the experts differ or agree on a matter. Here, the theoretical assumptions from organizational theory come to the fore again. There exists a debate amongst different academics about whether all theoretically deduced categories should be kept throughout the analysis or whether those categories that were not found in the empirical data can be deleted. The analysis at hand roughly applies and orients itself to Mayring's model; still, in case the analysis shows a discrepancy between theory and empiricism, this will be picked up and interpreted in the conclusion. In those instances, this thesis therefore takes up Gläser and Laudel's adaptations of Mayring's model (Gläser and Laudel 2004, p. 199).

To sum up, Step 1 has a more inductive approach, whereas Step 2 is essentially the deductive part of the coding. Step 3 considers both inductive and deductive empirical observations and puts these results into the overall theoretical context; this third step implies interpreting the new data received from the qualitative content analysis. This thesis thus refers to previous research on qualitative procedures, which also includes a mixed approach in its portfolio and gives the researcher the option to "use some combination of predetermined and emerging codes" (Creswell 2009, p. 187).

#### 4.3.2.1 Categorizing and Coding

The category system of this project represents the sum of all categories used during the qualitative content analysis (Kuckartz 2009, p. 337). In order to be able to code the textual data, the researcher needs a few rules for coding that he or she can apply to the text; the category system and rules for coding are subsumed as the codebook in quantitative research. The approach of this thesis, the mixture of deductive and inductive procedures, has already been explained above and also corresponds with recommendations in the social science literature (among others, Kuckartz 2009, p. 340). Concerning the categories for the analysis, these can be understood as "the instruments with which the text is worked through" (Mayring 2014, p. 37). As recommended by Mayring, it is helpful to list some examples for each category so that the coding is transparent to fellow scientists. Thereby, the researcher guarantees the intersubjective verifiability of his or her research project (Mayring 2014, p. 38). In this context, Mayring also responds to potential criticism on qualitative content analysis: in his opinion, it is necessary to refer to the latest research on the matter when making procedural decisions. And a researcher should always aim for theoretical stringency in his or her research, even if there might exist some technical fuzziness in the qualitative methods (Mayring 2014, p. 41). Mayring stresses that "content-related arguments should always be given preference over procedural arguments; validity is regarded more highly than reliability" (Mayring 2014, p. 6). As mentioned above, this thesis uses a combination of inductive and deductive approaches for category building. This methodology is consistent with Mayring, who introduces both approaches and—depending on the research project at hand—also gives the option of a combined inductive and deductive approach (Mayring 2014, p. 52). Figure 4.3 shows the theoretical assumptions of the guided interview and the respective coding for the interview transcripts.

#### 4.3.2.2 Interpreting and Analyzing

As mentioned above, this thesis broadly applies the model of Mayring for the qualitative content analysis. The choice for this model goes back to the fact that this analytical approach is perceived as the leading model when it comes to qualitative content analyses (Blatter et al. 2007, p. 76). Other scholars have already emphasized that Mayring specifically focuses on text analysis and that he presents various and context-specific techniques for using his model (Flick 2014, p. 429). Also, the way the model works—with a frame of coding and theory-led categories—means that this qualitative method is relatively systematic, as compared to other qualitative methods (Flick 2014, p. 430). The last advantage of Mayring's model is that it can be applied in a flexible manner, due to its deductive and inductive aspects. Nevertheless, the analysis at hand adjusts Mayring's model when it comes to the interpretation of relevant text passages; the underlying reason is that previous

(1) Assumption: The UNHCR's organizational environment influences its approach toward EDPs.
Coding: organizational environment

(2) Assumption: Inter-organizational relations influence the UNHCR's approach toward EDPs.
Coding: inter-organizational, competition, collaboration

(3) Assumption: The 'member states' have an influence on the organization's EDP approach.
Coding: member states, influence of states, state control

(4) Assumption: The UNHCR approaches the topic because it overlaps with the organization's goals.
Coding: organizational goals, inner motivation

(5) Assumption: The UNHCR's headquarters pushes the organization's EDP approach.
Coding: headquarter influence, field offices

(6) Assumption: Individuals in the organization push the UNHCR's EDP approach.
Coding: individual(s), leader(s),

(7) Assumption: The High Commissioner pushes the topic and has a personal motive for doing so.
Coding: High Commissioner, Guterres

(8) Assumption: The EDP approach is a means to enhance the UNHCR's public image.
Coding: organizational image, façade, propaganda

(9) Assumption: The UNHCR staff perceives the EDP approach as a long-term learning process.
Coding: staff learning, organizational learning

(10) Assumption: The UNHCR's EDP approach depicts a change in the organization.
Coding: organizational change, incremental change, critical moments, critical juncture

(11) Assumption: The UNHCR approaches the topic because it correlates with its norms / values.
Coding: organizational norms, organizational values, normative foundation

**Fig. 4.3** Theory-led assumptions on the empirical puzzle and the associated coding

research criticized that the aspect of interpreting the data has actually been neglected in Mayring's content analysis (Blatter et al. 2007, p. 84).

The analysis covers the time frame from 2005 until 2015. The underlying intention is to depict the central events and to show the possible changing organizational behavior. The analysis of the timeframe has a theory-led structure and thus is orientated toward the theoretical assumptions of this thesis (cf. Blatter et al. 2007, p. 182). Therefore, the analysis also keeps up with the theoretical focus and uses—once again—the theoretical assumptions developed in Sect. 4.2. In this context, Mayring also talks about a "structured content analysis" (Mayring 1991, p. 213), because certain aspects of the textual corpus are picked from the text corpus, and these paragraphs are organized according to the theoretical criteria of Sect. 3.3. Mayring stresses that this structured approach helps the researcher to interpret the empirical data; additionally, the previously developed theoretical categories guide the analysis and reduce the general text corpus into the relevant parts (Mayring 1991, p. 213). To reduce the text material is one of the central characteristics of the qualitative content analysis (Flick 2014, p. 429). Therefore, the rules for coding allow for the interpretation of the relevant paragraphs; the coding constitutes the preparatory step for the interpretation and analysis (Flick 2014, p. 375).

The researcher thus accomplishes the following steps during the qualitative content analysis by building upon Flick (2014, p. 421) and Mayring (2014, pp. 54–97): After I familiarize myself with the (primary and secondary) data, I code the interview transcripts based on my theoretical categories in order to find themes in the material (deductive approach). The deductive category assignment is guided by the rules for coding that briefly show what falls under which category and how a category can be identified in a text (Mayring 2014, p. 104). Following this, I scan the material once more to review and look for themes that arise out of the material itself (inductive approach), and I code these themes as well. Therefore, the analysis in Chap. 5 is a theory-oriented presentation of the results, expanding beyond a mere narrative depiction (Blatter et al. 2007, p. 183). Leaning on existing social science research, the analytical process can be described as a "theory-guided interpretative process" (Powell et al. 2012, p. 245).

To conclude, this chapter has presented the research design of this thesis. Firstly, the operationalization of security theory and organizational theory is essential to lay the theoretical groundwork for the analysis. The theoretical assumptions derived from these theoretical strands guide the qualitative expert interviews by means of a set of questions. Secondly, this chapter introduced the method of qualitative content analysis, including the way that the data will be categorized, coded and interpreted. The analysis in the subsequent chapter, Chap. 5, will be based on the aspects outlined in this chapter. In other words, the 'tools' of this chapter help to analyze the case in the next chapter.

## References

Biermann F, Siebenhüner B, Bauer S, Busch P-O, Campe S, Dingwerth K, Grothmann T, Marschinski R, Tarradell M (2009) Studying the influence of international bureaucracies: a conceptual framework. In: Biermann F, Siebenhüner B (eds) Managers of global change: the influence of international environmental bureaucracies. MIT, Massachusetts, pp 37–74

Blatter JK, Janning F, Wagemann C (2007) Qualitative Politikanalyse: Eine Einführung in Forschungsansätze und Methoden. VS Verlag für Sozialwissenschaften, Wiesbaden
Creswell JW (2009) Research design: qualitative, quantitative, and mixed methods approaches, 3rd edn. Sage, Los Angeles
Flick U (2014) An introduction to qualitative research, 5th edn. Sage, London
Gläser J, Laudel G (2004) Experteninterviews und qualitative Inhaltsanalyse als Instrumente rekonstruierender Untersuchungen. VS Verlag für Sozialwissenschaften, Wiesbaden
Kaiser R (2014) Qualitative Experteninterviews: Konzeptionelle Grundlagen und praktische Durchführung. Springer VS, Wiesbaden
Kratochwil F, Ruggie JG (1986) A state of the art on an art of the state. Int Organ 40:753–775
Kuckartz U (2009) Inhaltsanalyse. In: Westle B (ed) Methoden der Politikwissenschaft. Nomos, Baden-Baden, pp 334–344
Mayring P (1991) Qualitative Inhaltsanalyse. In: Flick U, von Kardoff E, Keupp H, von Rosenstiel L, Wolff S (eds) Handbuch qualitativer Forschung: Grundlagen, Konzepte, Methoden und Anwendungen. Beltz - Psychologie Verl. Union, München, pp 209–213
Mayring P (2014) Qualitative content analysis: theoretical foundation, basic procedures and software solution. SSOAR, Klagenfurt
Powell JJ, Bernhard N, Graf L (2012) The emergent European model in skill formation: comparing higher education and vocational training in the Bologna and Copenhagen processes. Sociol Educ 85:240–258
Westle B (2009) Teilstrukturierte und offene Befragungsformen. In: Westle B (ed) Methoden der Politikwissenschaft. Nomos, Baden-Baden, pp 248–259

# Chapter 5
# The UNHCR and Environmentally Displaced Persons

> *We speak about the humanitarian consequences of global climate change as though*
> *we are dealing with a future possibility. Far from it, this is not a possibility [. . .],*
> *this is a certainty. And this is not about the future, this is about now!*
>
> (Johnstone 2008a, p. 4)

The UNHCR's interest in questions concerning EDP protection is controversial: The organization is responsible for over 20 million refugees and has already taken up the topic of IDPs. Additionally, the UNHCR's mandate does not cover the protection of EDPs. Therefore, this analysis will tackle the question: Why, then, does the UNHCR approach the issue of EDPs the way that it does?

In this chapter, the empirical puzzle of the UNHCR's approach to EDPs will be considered. The analysis hereby builds upon the conceptual design developed in Chap. 4 and incorporates primary and secondary data on the UNHCR's EDP approach, including the findings of the qualitative expert interviews. In order to uncover the underlying impetus of the UNHCR's approach, the analysis is presented by means of a context-centered interpretation (the influence of the organizational environment, the influence of states, etc.) in chronological order.

## 5.1 The United Nations High Commissioner for Refugees

In order to get an overview of the organization, the question about the UNHCR's organizational structure arises. To start with, the mission of the UNHCR dates back to the beginning of the twentieth century. In 1921, the League of Nations founded a High Commissioner for Refugees who was in place until shortly after the Second World War. The first Commissioner for Refugees was Fridtjof Nansen (UNHCR 2010, p. 3). Following this first episode of refugee protection, the United Nations



S. Hantscher, *The UNHCR and Disaster Displacement in the 21st Century*,
Contributions to Political Science, https://doi.org/10.1007/978-3-030-19689-9_5

founded the International Refugee Organization (IRO) in 1946 which came with the specific mandate to repatriate Second World War refugees (UN Office of the High Commissioner for Human Rights 1993, p. 2). That said, refugee protection and stateless persons were some of the first matters tackled by the UN (UN Office of the High Commissioner for Human Rights 1993, p. 2).

However, a lot of states criticized the work of the IRO, its refugee definition being considered insufficient and the mandate inadequate for the refugee situation at hand (UN Office of the High Commissioner for Human Rights 1993, p. 3). This resulted in a lot of IRO membership fees unpaid by states. By the accounts of the Office of the United Nations High Commissioner for Human Rights (OHCHR), "it soon became evident that the responsibility for refugees deserved further international effort under the auspices of the United Nations itself" (UN Office of the High Commissioner for Human Rights 1993, p. 3). That is why, in 1949, the UN decided to found a successor to the IRO, the UNHCR, which was set up as a subsidiary organ to the UN General Assembly (UN GA) in 1951 (UN Office of the High Commissioner for Human Rights 1993, p. 3).

### *5.1.1 The UNHCR's Organizational Structure*

Within the UN, the UNHCR is part of a group of UN funds and programs—e.g., with the UNDP and UNICEF—that are characterized by voluntary funding (Dingwerth et al. 2011, p. 163; Pease 2016, pp. 15–43). These funds are often tied to specific purposes and are thus referred to as tied or earmarked aid. In order to raise money for its Annual Programme Budget, the UNHCR publishes a 'Global Appeal' in December and organizes a pledging conference every year (Betts et al. 2012, p. 97). During these annual conferences, donor states pledge to fund specific programs, but the UNHCR can never be sure of the whole budget being met or of all pledges being fulfilled. Adding to the Annual Programme Budget, the organization can raise money through special appeals to be able to act in cases of unforeseen events. Therefore, the "UNHCR must dedicate considerable time and effort to fundraising and donor relations" (Betts et al. 2012, p. 97). The top ten donor states of 2009 were the USA, the European Commission, Japan, Sweden, the Netherlands, Norway, the UK, Germany, Denmark, and Canada (Betts et al. 2012, p. 97). The composition of the top ten donors changed only slightly during the period under analysis. By 2015, the top ten donors were the USA, the UK, the European Union, Japan, Germany, Kuwait, Sweden, Norway, Denmark, and the Netherlands (UNHCR 2015).

As the tasks fulfilled by this group of UN programs are specialized and decentralized from the UN, the staff working at these organizations are experts on the topic and have often acquired earlier work experience outside the UN (Dijkzeul and Beigbeder 2003, p. 114). This also applies to the UNHCR: the organization consists of a High Commissioner and his or her staff who work at the headquarters in Geneva as well as in the field offices (Loescher and Milner 2011, p. 189). In other words, the UNHCR is distinctive because it is a bureaucracy that has its own organizational culture and value system and an individual, and both represent the UNHCR (Loescher et al. 2012, p. 74). As of 2018, the UNHCR's workforce reached 10,966 staff members disposed in 130 countries. More specifically, 87% of staff are based in the field (UNHCR 2017). As regards the UNHCR's role in the field, during the period of this analysis, some academics describe the organization's operational strength in refugee situations as being like that of a nation state, the refugee camps being the territory, the refugees constituting the citizens, and the UNHCR providing public services like water, health care and sanitation owing to an underlying ideology of gender equality and community participation (McConnachie 2014, p. 88).

Alongside the UN funds and programs of which the UNHCR is a part, there are specialized organizations such as the ILO, the UNESCO, the IMF, and the WHO (Dingwerth et al. 2011, p. 163). In respect of the structural classification inside the UN, it is relevant to outline the relationship between the UNHCR and other UN bodies. The UN GA and the United Nations Economic and Social Council (ECOSOC) govern the UNHCR and thus fulfill a policy-making function (UNHCR 2017). The organization's programs and budgets are approved by the UNHCR's Executive Committee (ExCom) (UNHCR 2017). Additionally, the UN GA appoints the High Commissioner. The High Commissioner is also responsible for presenting the organization's programs to the ECOSOC and to the UN GA in annual reports (UNHCR 2017). The organization's programs usually have a 2-year timeframe.

In the annual ExCom meetings in Geneva, the UNHCR, IOs, and NGOs take part in discussions concerning the organization's programs and budgets. Structure-wise, the ExCom is a subsidiary organ to the UN GA (UNHCR 2017). It was established in 1959 and is made up of member states which the ECOSOC elects "on the widest possible geographical basis from those states with a demonstrated interest in, and devotion to, the solution of the refugee problem" (Economic and Social Council, United Nations 1958). At the end of the period of this analysis, in 2015, the ExCom consisted of 98 states. The ExCom fulfills a variety of tasks which include "to advise the High Commissioner in the exercise of his/her functions, to review funds and programs, to authorize the High Commissioner to make appeals for funds and to approve proposed biennial budget targets" (UNHCR 2017). To fulfill the tasks of the ExCom throughout the year, the committee has a Standing Committee at its side which meets three times a year and thus keeps up the work of the ExCom. The overall direction of the organization is accounted for by the High Commissioner, the Deputy High Commissioner, and the Assistant High Commissioners for Protection and for Operations (UNHCR 2017). It is specified in the Refugee Convention that the High Commissioner shall fulfill his or her role in a non-political and humanitarian manner (UN Office of the High Commissioner for Human Rights 1993, p. 3). Filippo Grandi is the UNHCR's High Commissioner until 2020; his predecessor was António Guterres from 2005 to 2015.

### *5.1.2 The 1951 Refugee Convention*

Following this overview of the organization's structure, the question arises as to what is within the UNHCR's mandate and what issues lie outside it. The idea behind the establishment of the UNHCR was to have the organization oversee whether states comply with the right to asylum according to Article 14 of the Universal Declaration of Human Rights (Loescher and Milner 2011, p. 191) and in 2003, the UN GA unlimitedly extended the UNHCR's mandate (UNHCR 2017). The mandate of the UNHCR—the 1951 Convention Relating to the Status of Refugees (Refugee Convention)—is to guard the protection of refugees. The UNHCR's task is to guard the global refugee regime, that is, all the rules and principles that states decided upon to protect refugees (UNHCR 2010, p. 4). All of these written standards for refugee protection are obligations for states and are specified in the 1951 Convention. Originally, the scope of the Refugee Convention comprised only Europe and all kinds of refugee movement before 1951 (UNHCR 2010, p. 2). Additionally, the 1951 Convention defines who falls within the refugee category and what kind of rights come with being recognized as a refugee. A refugee is:

> Someone who is unable or unwilling to return to their country of origin owing to a well-founded fear of being persecuted for reasons of race, religion, nationality, membership of a particular social group, or political opinion (UNHCR 1951)

As the UNHCR is the principal organization in the refugee regime, it supervises that states accord with the 1951 Convention, the content of which is based on international law. In 1967, the Refugee Convention gained its global character with the adoption of the Protocol to the Convention (UNHCR 2010, p. 2). The role of the UNHCR in this regime is to see that refugees get access to protection and to monitor possible long-term solutions such as integration into a foreign country or reintegration into the home country (Loescher and Milner 2011, p. 192). Also, the Convention makes statements on the applicable civil and economic rights, the right to education and employment, and the right of access to national courts for refugees. Additionally, the Convention includes the principles of non-refoulement (Article 33), non-penalization, and non-discrimination (Article 3), which show the rights-based character of the Convention. The first principle, non-refoulement, provides that states may not return a refugee to a territory where the person is threatened; the second principle, non-penalization, affirms that states may not charge refugees with criminal or immigration offences (even when refugees breach immigration rules) or detain asylum-seeking refugees; and the third principle, non-discrimination, declares that states shall apply the Convention "without discrimination as to sex, age, disability, sexuality, or other prohibited grounds of discrimination" (UNHCR 2010, p. 3). For the period of this analysis (2005–2015), 148 states were parties to either the 1951 Convention or the 1967 Protocol or both (UNHCR 2015).

Concerning its global impact, the Refugee Convention experienced some regional expansions. The Organization of African Unity (OAU) defines a refugee as fleeing "external aggression, internal civil strife, or events seriously disturbing public order" (Organisation of African Unity 1969, p. 2) as stipulated in the

Convention Governing the Specific Aspects of Refugee Problems in Africa. In this context, the OAU Convention does not duplicate but rather complements the Refugee Convention (UN Office of the High Commissioner for Human Rights 1993, p. 5). Additionally, the Central American declaration—the so-called Cartagena Declaration of 1984—also incorporates other causes of flight that make for refugee status, namely those circumstances that seriously disturb public order (Loescher and Milner 2011, p. 191). Nevertheless, "even in the regional contexts of Africa and Central America, cross-border displacement in the majority of climate-related cases will not be covered by refugee law" (Kälin 2010, p. 89). In addition to these regional expansions, the matter of refugee protection is also part of international law: the international human rights norms intend to help those individuals who are unable to provide the necessary evidence to fulfill the 1951 Refugee Convention (Loescher and Milner 2011, p. 191). Being confronted with these diverse realities and various causes of flight in the last decades, the UNHCR followed a rather flexible interpretation of the Refugee Convention, even though states did not always appreciate this step (Loescher and Milner 2011, p. 192).

### *5.1.3 Mandate Changes at the UNHCR*

The UNHCR does not exist in a vacuum and the changing international scene since its establishment after the Second World War hugely affected how the organization currently interprets its mandate. During its early years of existence, the UNHCR mainly focused, first, on refugees from the Second World War and, second, on the people who fled from the communist countries of Eastern Europe. In the following decades, in the 1960s and 1970s, the organization widened its scope and the focus shifted to include questions of refugee protection in the global South that arose due to decolonization trends.

The most cited examples of mandate change are the questions of repatriation and of internally displaced persons (IDPs). Concerning the question of repatriation, the UNHCR had pushed repatriation as a possible addition to host country integration or as a means of asylum in a third country. Particularly before 1980, the UNHCR's approach implied that "repatriation had to be both safe and voluntary because forced repatriation violates the international legal principle of nonrefoulement, which is the cornerstone of international refugee law" (Barnett and Finnemore 1999, p. 722). Nevertheless, the UNHCR's references and the ExCom resolutions show that the repatriation criteria—to be safe and voluntary—blurred over time and consequently led to "a lowering of the barriers to repatriation and an increase in the frequency of involuntary repatriation" (Barnett and Finnemore 1999, p. 722).

On the question of IDPs, the UNHCR opened in terms of approaching this topic and helping on the ground (Dingwerth et al. 2014, p. 186). The organization's concern for IDPs led to a UN GA decision (General Assembly 1993) that supports the UNHCR's efforts to protect IDPs and that preceded an ExCom conclusion (General Assembly 1994) that also encouraged the UNHCR's involvement in the matter of IDPs. This development shows that the organization has adapted to new situations in the field offices and has tried to avoid giving the impression of a static and inflexible organization. More specifically, the matter of IDP protection directly touches the issue of national sovereignty, which makes the UNHCR's extension into this matter even more exceptional. Overall, since the 1980s, the UNHCR has shown a tendency to advance its role in development policies and in actions that prevent refugee flows (Schorr 2004, p. 62).

The influence of states on the UNHCR's decision-making shows in the UN GA's role (and, to a lesser extent, that of the ECOSOC's) to amend the UNHCR's mandate (Türk 1999, p. 154). As both the UN GA and the ECOSOC have amended the mandate in the past, the actions of the UNHCR exceed its original statute of 1951, which shows in the resolutions from both bodies and in the implied powers and de facto practices of the UNHCR (Türk 1999, p. 154). The legal basis of the UN GA's influence goes back to Article 22 of the UN Charter which introduces the UNHCR as a subsidiary organ of the UN GA. Nevertheless, the legal basis for the UNHCR has not changed since the end of the Second World War. Concerning the members of the UNHCR, the organization characterizes itself as having undergone a rise in member states and therewith a rise in heterogeneity amongst its members since 1979 (Dingwerth et al. 2014, p. 194). More specifically, the state representatives of developing countries in the UNHCR's ExCom are on the rise compared to developed countries (Dingwerth et al. 2014, p. 195). The UNHCR budget largely consists of contributions from states: only a small proportion—approximately 2–5% of the UNHCR's budget—comes from the UN and this has proportionately declined over the years (Hüfner 2015, p. 110). As a result, the organization appeals to donors for voluntary contributions to its annual budget or to its ad hoc special appeals (Betts 2009, p. 16). The rise in the UNHCR's budget over the years, and especially in the last decade, thus goes back to the UNHCR's successful fundraising campaigns (Dingwerth et al. 2014, p. 198). Here, the question of the financial contribution of states is directly related to the question of burden-sharing: whereas the receiving countries—i.e., the developing countries—advocate involving the local population in budget planning, the big donor countries—i.e., the developed countries—sometimes demand the assignment of developing tasks to be fulfilled by development organizations such as the UNDP or the World Bank (Dingwerth et al. 2014, pp. 200–201). In answer to donor states' call for financial transparency inside the UNHCR, the organization introduced the 'Global Reports', a detailed yearly report that, since 1998, depicts all programs and expenses.

## 5.2 Analysis of the UNHCR's Approach Toward EDPs

The following four sections correspond to the theory-led assumptions developed in Chap. 4 and, by this means, present the findings of the empirical analysis. The analysis also presents a new assumption that arose throughout the data collection phase and its addition to the UNHCR's interest in EDPs.

### *5.2.1 The UNHCR's Organizational Environment*

This first section presents the knowledge gained from the organizational environment and surroundings, the overall assumption being that the external circumstances and the organizational environment have an impact on the internal circumstances of the organization (for a summary of this section see Table 5.1). The following paragraphs present the insights gained from the primary data coding of 'hazard', 'inter-organizational environment', and 'organizational environment' and puts these results into context with the information gained from secondary data. The code 'hazard' arose during the coding of the material, whereas the other two assumptions were developed prior to the field research in a theory-led manner (see Sect. 4.2.1). As a short recapitulation of Chap. 4, the following theory-led assumptions lead this section on the organizational environment:

(1) The UNHCR's organizational environment influences its approach toward EDPs.
(2) Inter-organizational relations influence the UNHCR's approach toward EDPs.

(New assumption): The occurrence of hazards influences the approach to EDPs.

#### 5.2.1.1 The Influence of the Organizational Environment

The focus will first lie on the assumption that the overall organizational environment has an impact on the UNHCR's decisions concerning EDPs.

(1) The UNHCR's organizational environment influences its approach toward EDPs.

**International Actors**

On the topic of environmental migration and displacement, IOs such as the UNHCR are important actors. The actors working on the issue range from the UNHCR to the IOM and the ILO, as well as the OHCHR (on topics such as human rights of migrants), the United Nations Population Fund (on topics such as demography), and the World Bank (on topics like economic growth and circular migration) (Betts 2011, p. 16). Also, various global actors monitor the access to resources and mediate vulnerability in cases of environmental catastrophes, very often down to the local level (Mearns and Norton 2010, p. 65). The various spheres of governance of EDPs lead to a diversity of actors who engage in the issue (see Sect. 1.1).

**Table 5.1** The influence of the UNHCR's organizational environment

| Ideational depiction<br>(broadly informed by security theory) | Organizational factors<br>(operationalized organizational theory) |
|---|---|
| **2005–2011** | |
| – Since the **2000s**: Rise in the number of publications on the topic<br>– **2007**: UN SC discusses climate change for the first time<br>– Advocacy standpoint: To provide evidence about future developments<br>– **2009**: Rise in alarmist (securitizing) publications; active advocacy work of international actors, including the UNHCR<br>– From an advocacy standpoint, climate change is a better trigger than natural disasters<br>– **2009**: UN Secretary General report on climate change and possible security implications<br>– **2009**: UN GA resolution on climate change and security | – **2004**: Tsunami in Southeast Asia triggers a call to understand displacement in disaster settings; disaster focus of the international community<br>– **2005**: UNHCR becomes the inter-agency cluster lead for protection and responsibility for conflict IDPs at the country level<br>– **2005–2015**: OHCHR, UNICEF, and UNHCR share responsibility for people displaced by natural disasters in the Global Protection Cluster; no designated lead agency for the field level<br>– **2008**: IOM begins to work on the environment-migration nexus; becomes a forerunner on the topic<br>– **2008**: Cooperation in the IASC informal task force on migration and displacement (IOM, OCHA, UNHCR, etc.)<br>– Since **2009**: International rise of interest in climate change; shift from a disaster-centered concept to climate change thinking<br>– **Pre-COP 15**: UNHCR and other IOs position themselves on the topic; UNHCR and IOM collaborate but also compete in their advocacy |
| – **Conclusion A:** The topic of EDPs emanated from the international environment and was picked up by the UNHCR<br>– **Conclusion B:** The disaster-thinking from 2004 to 2006 had developed into a climate change perspective by 2009<br>– **Conclusion C:** Differences in how many staff the IOs assigned to the topic | |
| **2011–2015** | |
| – **Advisory Group**: Pushes for a human security-centered perspective on EDPs in the international forums; compliments the UNFCCC for addressing human security in its 2014 report<br>– **2014**: Policy brief of the UNU and the Nansen Initiative has a human security depiction<br>– **2014**: San Remo Consultation background document shows a strong human-centered perspective; climate change is looked at from the individual's perspective<br>– **2015**: Advisory Group publication pre-COP 21 has a human security focus | – **2010s**: War in Syria and Iraq and refugee movement to Europe catches a lot of state attention and ties up many resources<br>– **2010**: Breakthrough at COP 16 (Art. 14(f)); Cancun Declaration is the external impetus that initiates the Nansen Initiative; IOs succeed in pushing the topic (leading to Sendai Framework and Paris Declaration)<br>– **2011**: 'IASC Operational Guidelines' guide the UNHCR's work on natural disasters<br>– **Feb. 2011**: Inter-organizational collaboration at the expert roundtable on climate change and displacement in Bellagio, Italy<br>– **June 2011**: ExCom Standing Committee discusses the UNHCR's role in natural |

(continued)

**Table 5.1** (continued)

| Ideational depiction<br>(broadly informed by security theory) | Organizational factors<br>(operationalized organizational theory) |
|---|---|
| | disasters; the UNHCR should only complement state responsibility<br>– IO competition: All try to enhance visibility, reputation, impact, and fundraising potential<br>– **Dec. 2011**: UNHCR Ministerial Conference (see Sect. 5.2.2); states are reluctant concerning the UNHCR's EDP approach; birth of the Nansen Initiative; UNHCR positions itself strategically as an observer in the steering group<br>– **2012:** Nansen Initiative is the first step of the international community to approach EDPs<br>– IOM and UNHCR cooperate extensively in the advisory group<br>– **2012**: COP 17 (Doha) (Art. 7(a)(vi))<br>– **2014**: San Remo Consultations by the UNHCR, Brookings Institution, and Georgetown University; planned relocation, migration, and displacement are the types of human mobility to be considered by climate change adaptation measures<br>– UNHCR has leading role in the Advisory Group on Climate Change and Human Mobility<br>– Debates whether it needs a separate IO for the EDP protection gap; states are reluctant because they anticipate a fight for funding among IOs<br>– **2015**: COP 21 (Paris): side events of the UNHCR focus on planned relocation<br>– **2015**: Protection Agenda adopted by 114 state delegates; Platform on Disaster Displacement resolves some of the struggles with terminology<br>– Migration is only tackled as a side issue at the UN until 2016 |
| – **Conclusion A:** Deliberations on how to establish EDPs as a topic in the UN; UNHCR and IOM pushed the topic, otherwise it would not have been on the agenda at all<br>– **Conclusion B:** War in Syria and Iraq and refugee movement to Europe caught a lot of state attention and tied up many resources<br>– **Conclusion C:** No organizational interest at the UNHCR before 2009; the international community's focus on climate change and natural disasters acted as an external impetus | |

An example of how the organizational environment influenced the UNHCR is the UN's Inter-Agency Standing Committee (IASC). The IASC was established in 1992 and is a forum that involves UN- and non-UN humanitarian partners. As established in UN GA Resolution 48/57, the IASC is the leading mechanism when it comes to the inter-agency coordination of humanitarian assistance. In the IASC, key partners work together and develop policies and take part in decision-making. The IASC is led by the Under-Secretary General and Emergency Relief Coordinator (USG/ERC) of OCHA. In 2005, the IASC handed over responsibility to the UNHCR to protect conflict-generated IDPs, to provide emergency shelter, and to coordinate and manage IDP camps (Goodwin-Gill and McAdam 2017, p. 10). The responsibility for people displaced by natural disasters was shared by OHCHR, UNICEF, and the UNHCR in the Global Protection Cluster. In 2006, the IASC adopted the 'Operational Guidelines on Human Rights and Natural Disasters' to guarantee the human rights of people who suffer from natural disasters, whether the people had been displaced or whether they were able to remain in their homes (Kälin 2010, p. 93). As a follow-up, the IASC published the 'Operational Guidelines on the Protection of Persons in Situations of Natural Disasters' in 2011. These guidelines are the basis for all humanitarian actors contributing to the humanitarian response after a natural disaster. All international humanitarian organizations—and therefore also the UNHCR—"are guided by these Operational Guidelines in their activities, in particular when monitoring and assessing the situation and needs of affected persons" (Inter-Agency Standing Committee 2011, p. 21). That said, the IASC's decision-making and guidelines directly impact the work of the UNHCR.

**International Initiatives on the Topic**

The following developments and international initiatives in the organizational environment of the UNHCR fall into the period of analysis of this thesis. Relating to the assumption introduced above, the UNHCR's effort on the issue of EDPs increased at a time when the topic of EDPs was on the rise. Particularly with regard to events in international forums, the relevance of EDPs increased in the eyes of many in Geneva.

First, the UN Security Council (UN SC) discussed the topic of climate change for the first time in 2007. The United Kingdom of Great Britain and Northern Ireland, being the president of the UN SC in 2007, declared that "an unstable climate will exacerbate some of the core drivers of conflict, such as migratory pressures and competition of resources. The 'Stern Review Report on the Economics of Climate Change' speaks of potential economic disruption on the scale of the two world wars and of the Great Depression. That alone will inevitably have an impact on the security of all of us—developed and developing countries alike" (United Nations Security Council 2007, p. 2). At the same UN SC meeting, the then UN Secretary General, Ban Ki-moon, took the floor and warned that "migration driven by factors such as climate change could deepen tensions and conflicts, particularly in regions with large numbers of internally displaced persons and refugees" (United Nations Security Council 2007, p. 14). On behalf of the European Union, the German delegate warned that "there is a clear link between climate change and the need for conflict preventions" (United Nations Security Council 2007, p. 19).

Second, states took the floor at various UN GA meetings to outline their standpoint on the topic. In 2009, the European Union declared that it "sees climate change as a universal threat that will create new security dynamics and risks in all regions—

between States and within States—by intensifying conflicts over depleting or scarce resources, by causing economic damage and adding pressure to critical infrastructure, and also by amplifying or triggering environmentally induced migration, thus increasing pressure on international governance" (UN General Assembly 2009a, p. 5). In September 2009, at the request of the UN GA member states, the UN Secretary General prepared the report 'Climate Change and its Possible Security Implications' (UN General Assembly 2009b), shortly before the Conference of Parties (COP 15) of the United Nations Framework Convention on Climate Change (UNFCCC) in Copenhagen. The UN GA resolution on climate change and security of 2009 reads as follows: "Deeply concerned that the adverse impacts of climate change, including sea-level rise, could have possible security implications. 1. Invites the relevant organs of the United Nations, as appropriate and within their respective mandates, to intensify their efforts in considering and addressing climate change, including its possible security implications" (UN General Assembly 2009c, p. 2). Even though it is clearly stated that the respective UN organ shall act within its mandate, this call was not unheard by the UNHCR.

Third, one prime example of these influential events during the 2000s was the COP in Copenhagen in 2009.[1] As one NGO staff member describes, "we had responded to it [environmental displacement] earlier but more ad hoc and not really part of our strategy. And people tended to read displacement only in conflict displacement terms" (NGOR1, para. 2). At the time, the focus on conflict displacement was still very much present in many IOs and NGOs and most of the money was channeled into the conflict work (NGOR1, para. 7). The intention of NGO and IO representatives was to get environmental displacement on the international agenda and to get the topic into Copenhagen. Both the rise in alarming publications on EDPs and the active advocacy work of international actors shortly before Copenhagen go back to this underlying intention to have the topic acknowledged in the Copenhagen declaration (NGOR1, para. 2). According to one expert, "there was a big push toward Copenhagen" (NGOR1, para. 2) and in the months leading up to the COP in Copenhagen, various organizations like the IDMC, the NRC, the IOM, and the UNHCR began to position themselves on the topic of climate change and human mobility. As one expert admits, there was slight competition among the IOs and NGOs on the question of who does the best advocacy on environmental displacement and which organization(s) will occupy the topic post-Copenhagen (NGOR1, para. 5). One expert recalls: "For [organization X] it was clear in Copenhagen that unless [organization X] takes it on, there will be someone else competing for it. And it doesn't make sense to have two organizations on this. If you look at the reality, this is where you have the big numbers. So, if you want to secure funding for activities, this is also where the money will go. If you did your work right" (NGOR1, para. 5).

Against the background of the COP in Copenhagen, it becomes apparent that the subject of environmental displacement emanated from the international

[1]For the impact of the COP in Copenhagen on global climate governance, see McGee and Steffek (2016).

environment. In other words, "what is important to see is that from an advocacy point of view, you will pick something up from the field offices, but on climate displacement there has been a drive from the global" (NGOR1, para. 6). Additionally, the 2006 focus on a disaster concept slowly changed into a climate change perspective leading up to the COP in Copenhagen: "There was a growing knowledge that climate change will also create climate refugees and these big disasters, they are not that often. They only happen once or twice a year. But they kind of tend to fade away. Whereas climate change is constantly unresolved; it is kind of growing up!" (IOR3, para. 17).

Fourth, even though the COP in Copenhagen in 2009 did not measure up to its expectations, the COP in Cancun in 2010 was a breakthrough: a collaboration of IOs, NGOs, and committed states advocated the topic of environmental displacement, which resulted in paragraph 14(f) of the final declaration: "Measures to enhance understanding, coordination and cooperation with regard to climate change induced displacement, migration and planned relocation, where appropriate, at the national, regional and international levels" (UNFCCC 2011). According to IO representatives, the establishment of paragraph 14(f) in the agreement was a huge success and it led states to rethink their position on displacement and climate change (IOR5, para. 71–72). Therefore, the Cancun Declaration can be interpreted as the external impetus that initiated the Nansen Initiative (see below) (IOR5, para. 99–100). Added to this, the Doha Climate Change Conference (COP 18) in 2012 is another cornerstone: the international actors and COP participants repeatedly mention the following paragraph 7(a)(vi) in the decision on loss and damage as a step toward more international awareness about human mobility: "Acknowledges the further work to enhance the understanding of how impacts of climate change are affecting patterns of migration, displacement and human mobility" (Franck 2015, p. 10).

Fifth, post-2009, the issue appeared in global forums—e.g., at the COPs, in EU projects, and most noticeably in the Nansen Initiative. The Nansen Initiative goes back to the UNHCR Ministerial Conference in December 2011 during which Norway and Switzerland pledged to work together with several states and IOs on the matter of cross-border displacement in the context of disasters and the effects of climate change (The Nansen Initiative 2017). It consists of a steering group comprising Switzerland, Norway, Germany, Australia, Bangladesh, the Philippines, Kenya, Costa Rica, and Mexico. The Nansen Initiative Prediction Agenda lists the UNHCR as one of the committing partners, which shows that the UNHCR was able to jump on the bandwagon and position itself strategically in this newly developing sphere of EDP involvement. The UNHCR was an observer in the steering group of the Nansen Initiative (Türk et al. 2015, p. 9). The Nansen Initiative is perceived as a first step of the international community to approach the matter of EDPs. Since its start in 2012, the initiative has gathered all existing facts on EDPs and looked into the situation in different regions of the world (SR3, para. 14). As one state representative declares, the "Nansen Initiative has given us several ideas about how to deal with this [EDPs]. One of the main outcomes is that there are a lot of tools to deal with these situations. But those tools could be different from one region to another and that has to do with the differences among the regions" (SR4, para. 14). The main characteristic of the Nansen Initiative is that it is a state-led process. Many states saw this as essential and as a precondition for their support (SR5, para. 8). Additionally, a lot of IO representatives supported the initiative because it was a bottom-up process, including regional consultations in almost all regions of the world and assisted by civil society, academics, regional organizations, IOs, and states (SR5, para. 8). The Nansen Initiative consultation had a broad structure: it was a multi-stakeholder process including regional consultations. During the writing of the final report, around 50 experts (individuals and organizations) worked together in a consultative committee and the final product includes various opinions and views on the matter. One expert described the consultation as a "very inclusive and bottom-up process" (SR5, para. 8). The people who worked at the Nansen secretariat from the different organizations were only seconded to the Nansen for 60–70% of their workload; thus, they kept their job at their home IO. This structural idea goes back to one of the organization's experts, who argued that this structure allows the Nansen staff to be part of both the Nansen Initiative and their home IO (NGOR1, para. 31).

Sixth, and adding to the Nansen Initiative, the Advisory Group on Climate Change and Human Mobility is a very active forum and referred to by many of the experts when talking about EDPs. With about 100 members, the advisory group consists of academics (Sciences Po-CERI), a variety of IOs (NRC, IDMC, Refugees International, Arab Network for Environment and Development), members of the UN family (UNHCR, IOM, UNU-EHS, UNDP), and NGOs. During the period of analysis, Marine Franck of the UNHCR was the advisory group's coordinator (Advisory Group on Climate Change and Human Mobility 2015, p. 9). The advisory group's agenda is to "share the most recent available knowledge"; it "provides technical support to UNFCCC Parties and ensures that human mobility is taken into consideration in global climate negotiations" (Türk et al. 2015, p. 12). The underlying intention of the advisory group is that state of the art information filters into the steering group and, from there, into civil society initiatives and the approaches of private companies (SR3, para. 47). Ultimately, the high-level members of the steering group are kept updated about the empirical news and forecasts. During the debates at the advisory group, "we had deliberately agreed to not put locks on things, because as soon as you put locks on things, you need to have it signed off. This makes it bureaucratic and results in sensitivities among organizations" (NGOR1, para. 33). The advisory group's aim was to keep the various partners in the discussions, and the group therefore lowered the stakes for some of the participants (NGOR1, para. 33). Also, the advisory group organized press conferences and events and made submissions to the UNFCCC (Türk et al. 2015, p. 12). What is clear is that the advisory group strengthened the human security-centered perspective on EDPs in the international forum. In its 'Joint Submission to the UNFCCC on National Adaptation Plans' (NAPs) in July 2014, the advisory group compliments the Intergovernmental Panel on Climate Change (IPCC) on addressing human security for the first time in one of its chapters in its 2014 report (Advisory Group on Climate Change and Human Mobility 2014, p. 2). That said, the advisory group's participants concurred with this human security depiction instead of a securitization of the topic.

The end of the timeframe of analysis stands out because it overlaps with the end of the Nansen Initiative and the start of the follow-up, the Platform on Disaster Displacement. At an intergovernmental global consultation event in October 2015, 114 state delegates adopted the Agenda for the Protection of Cross-Border Displaced Persons in the Context of Disasters and Climate Change, also called the Protection Agenda. On the international level, the follow-up of the Nansen Initiative also played an important role. The Platform on Disaster Displacement goes back to July 2016 and Germany and Bangladesh chaired the platform for the first couple of years. Even though this second phase of the Nansen Initiative falls outside the period under observation here, it is necessary to acknowledge that the Nansen Initiative did not end in talks; instead, the wording of a 'platform' of this second phase makes clear that the second phase will be all about implementation (SR5, para. 10).

**International Events that Impacted the UNHCR**

Adding to these structural and political aspects of the international environment that had an influence on the UNHCR's decision-making, international politics also exerted an influence on how the UNHCR acted until 2015. International crises had an impact on how the international community of states and IOs handled the topic of EDPs. In this context, the experts in Geneva repeatedly mentioned the conflicts in Syria and Iraq as important events to factor into the analysis (AR1, para. 36; SR3, para. 8; SR5, para. 12). More specifically, the Syrian conflict and the rise of ISIS explain why international awareness on the matter of EDPs has declined (SR3, para. 8). The occurrence of conflict refugees therefore overshadowed the existence of EDPs, and today's crises slow down commitment to future crises due to environmental changes. Additionally, migration to Europe in the 2010s caught a lot of state attention and tied up many of the much-needed resources. As one state representative specified, one could observe an "absorption of attention" (SR5, para. 12), and even though the migration to Europe and the movement of the environmentally displaced are two faces of the same coin—i.e., events of migration and displacement—the topic of EDPs was not picked up and did not echo during the 'wave of refugees' to Europe in the 2010s. Rather, "everyone talks about the European crisis and the Syrian crisis and the African and Mediterranean migration routes. Other topics have difficulties making the international community aware of their existence" (SR3, para. 8). Therefore, the experts in Geneva mention the Syrian crisis as one reason that the UNHCR's approach at the ExCom in 2011 failed (see Sect. 5.2.2): the political crises in Syria and Iraq absorbed all forces and that is one reason for the reluctance of states concerning the UNHCR's involvement with EDPs (AR1, para. 36 and 39).

Many of the experts in Geneva concur in their assessment that, even though the context has changed, some of the instruments that are applied today are 70 years old (SR4, para. 13). According to one expert, "the conventions are a photograph of the time; they are a snapshot which reflects the critical situation in that moment" (SR4, para. 14). One state representative summarizes this lack of up-to-datedness of the international system as follows: "What about when we are trying to establish a strategy? When we talk about strategy, the UN system only talks to the UN system" (SR4, para. 19). Hence, the UN system as a whole was not and is still not capable of officially incorporating the topic of EDPs into its system (SR1, para. 50–51). In other words, the UNHCR's approach to EDPs was triggered by the feeling that the UNHCR's surroundings—i.e., international forums like the UN—were not able to act on the issue. IOs such as the IOM and the UNHCR had to evolve and needed to move toward the EDP topic, otherwise the issue would not have been on the agenda at all. IO and state representatives debated (or rather, still debate) the question of whether it needs an IO that only deals with climate change-induced migration. A lot of states do not see this as an attractive solution to the protection gap because the new organization would have to fight for funding just as much as the existing IOs. Nevertheless, some state representatives approve this idea and support a stronger UN focus on migration that comes along with it (SR1, para. 48). What goes hand in hand with this is that even though "most of the UN organizations deal with migration in one way or another, they do not occupy themselves with migration; they tackle migration as a side issue" (SR4, para. 10). One state representative sums up the global approach to EDPs up to 2015: "There were some agencies that dealt with migration, such as the IOM, that was not part of the UN system. We dealt with refugees with the UNHCR. We dealt with the Global Forum of Migration and Development which is also aside of the UN system. It is a parallel part" (SR4, para. 10). The incorporation of the IOM as an organization related to the UN in September 2016 can thus be interpreted as a result of this discontent with the situation up to 2015.

#### 5.2.1.2 The Influence of Inter-organizational Relations

The organizational environment of the UNHCR also includes other IOs and NGOs. Therefore, this analysis also captures the inter-organizational environment of the UNHCR on the matter of EDPs, working with the following theory-led assumption:

> (2) Inter-organizational relations influence the UNHCR's approach toward EDPs.

The results of the empirical analysis will be presented by incorporating both the collaborative and the competitive behavior of the IOs working on the topic of EDPs.

**Organizational Collaboration**

In the international arena, there exists only a small number of actors who work on matters of international migration, refugees, and displacement. As depicted in Sect. 1.1, these organizations all have a unique focus and cover a specific subject area. The UNHCR's inter-agency collaboration on the topic goes back to the first collaborations with the IOM in the 1990s and the accompanying 1996 publication of 'Environmentally-Induced Population Displacements and Environmental Impacts resulting from Mass Migrations' (UNHCR et al. 1996). As the title suggests, the focus lay on both EDPs and the impact on the environment of, for example, large refugee camps. Articles on this report mention that inter-agency collaboration on the topic is essential and that "professional rivalry must not be allowed to exist" (Kimani 1996, p. 119).

When it comes to EDPs, the IOM plays a significant role in international forums, particularly in terms of pushing the issue. The IOM was founded in 1951 and is a project-based organization (IOR3, para. 49–50). Since the early 2000s, it has repeatedly published on the topic of EDPs and in 2008 the IOM began to work on the climate-environment-migration nexus (SR2, para. 12). Hence, the IOM was the first organization that extensively considered the topic, before the UNHCR and before most NGOs (SR2, para. 12). At that time, the only information available was on climate change adaptation but it did not take human mobility into consideration. As a result, the IASC mandated an informal task force in 2008 which was convened by the IOM but with the collaboration of the UNHCR, OCHA, and other international actors (Informal Group on Migration/Displacement of the IASC 2008). This early work represents the first steps of the UNHCR-IOM collaboration on the topic of EDPs.

Between 2014 and 2017, the IOM cooperated with the UN University and other research partners in a project called Migration, Environment and Climate Change: Evidence for Policy (MECLEP). The MECLEP project resulted from this early academic research and still keeps the IOM in the inter-organizational network as a knowledgeable player. The organization was able to strengthen its efforts on the topic and has established a division that solely works on environment-related displacement and migration. The IOM's organizational focus on the topic has been followed and respected by states (SR1, para. 25; SR4, para. 17–18; SR4, para. 21–22). Also, from an early stage, the IOM implemented knowledge from extensive research into its field office programs (NGOR2, para. 57). The underlying argument of the IOM is that its staff already have to deal with the topic of displacement resulting from various triggers and will have to do so even more in the future. By restructuring its programs in the field offices and by guaranteeing that the up-to-date information trickles down to the field offices, the IOM has been described by its partners as very systematic and principled on the matter (NGOR2, para. 57). It is an open secret in Geneva that the reason this organization has been proactive on the issue is to defend its unique selling point when it comes to all matters of migration, including displacement situations (NGOR2, para. 57). By the same token, the actors collaborating with the IOM acknowledge its role in the issue due to its scientific data and expertise (NGOR2, para. 63). Against the background of the lack of an international migration regime (as compared to the refugee regime), as well as the absence of a migration organization in the UN family until 2016 and the nonexistence of an elaborate legal basis for matters of climate change migration and disaster displacement, the IOM tried to fill these gaps with its section on environmental migration and its climate change expertise.

However, the UNHCR has also identified these gaps and officially follows the rule that all agencies should address parts of the issue (IOR5, para. 33–34) and that the different agencies should focus on different aspects of the issue (IOR5, para. 38). Inside the UN family, the UNHCR was given the role of leading protection issues in complex emergencies. Since 2005, the organization has been the inter-agency cluster lead for protection at the global level; at the country level, the UNHCR was the lead in situations of conflict-induced displacement (Executive Committee of the High Commissioner's Programme 2011, p. 2). In cases of natural disaster, however, the UN did not designate a corresponding lead agency at the field level between 2005 and 2015 (UNHCR 2012, p. 4). Instead, to fill this protection gap, the UN decided to give its three protection agencies the responsibility to determine which agency is best-placed to lead in a specific emergency: the UNHCR, UNICEF, or the Office of the OHCHR (UNHCR 2012, p. 4).

Furthermore, in 2009, the UNHCR and the IOM collaborated repeatedly by submitting requests to the UNFCCC (Advisory Group on Climate Change and Human Mobility 2015; UNHCR et al. 2009a, b). Also, the UNHCR collaborated with the NRC, the Representative of the Secretary General on the Human Rights of Internally Displaced Persons, and the UN University (UNU) by submitting papers to the UNFCCC's Ad Hoc Working Group on Long-Term Cooperative Action under the Convention (UNHCR et al. 2009c). Here, the close cooperation between the different IOs shows in the organizational working relations before the COP in Copenhagen in 2009. Against the background of both the high impact hazards of the 2000s and the implications of climate change, the IOs decided that the 'business as usual' approach had to stop (AR1, para. 31). Instead, the international approach had to change on two levels: "first, states need to take on responsibility for hazards and the following humanitarian consequences [...]. And second, it requires more money" (AR1, para. 31). This remark perfectly sums up the reasons for the extensive inter-organizational cooperation on the topic of EDPs. From the 2010s onwards, the UNHCR and the IOM cooperated in the Advisory Group on Human Mobility and Climate Change, which was led by the UNHCR.

Nevertheless, the IASC, governments, and NGOs criticized the existing consultative process between the UNHCR, UNICEF, and the OHCHR of the 2000s (Executive Committee of the High Commissioner's Programme 2011, p. 2). More specifically, the "delays in designating a protection-cluster lead may affect the extent to which protection concerns are reflected in needs assessments [...] resulting in the denial of protection and assistance to those in need" (Executive Committee of the High Commissioner's Programme 2011, p. 2). In the words of the UNHCR, this collaboration inside the UN family "has led to delays and unpredictability. The High Commissioner has expressed willingness to take on a more predictable role" (UNHCR 2012, p. 4). This led to the IASC's proposal in early 2011 to pilot a new arrangement for the UNHCR: the organization should adopt a more predictable leadership role in the protection cluster in cases of natural disaster. The IASC's pilot arrangement proposed by the committee in January 2011 specifically focused on which humanitarian organization was responsible for the cluster lead at the country level. On condition that a state affected by a natural disaster requests UN support, the IASC suggests the following arrangement: "Where a protection coordination mechanism already exists, that arrangement and its leadership will continue.

Where OHCHR is present with adequate capacity and expertise and UNHCR is not, OHCHR will assume the protection coordination role. Only where these two elements are not present will UNHCR assume the role of leading and coordinating the protection cluster" (Executive Committee of the High Commissioner's Programme 2011, p. 7).

Collaboration at the inter-organizational level also showed in February 2011 during the expert roundtable on climate change and displacement that took place in Bellagio, Italy, and was organized by the UNHCR. The final summarizing document states the UNHCR's role among the other international actors as follows: "It was acknowledged that UNHCR has particular expertise in the protection dimensions of displacement within the framework of the global cluster approach and the Inter-Agency Standing Committee and in collaboration with other international organizations" (UNHCR 2011, p. 6). As a result, the ExCom's Standing Committee meeting in June 2011 discussed the matter of UNHCR's engagement with regard to people affected by natural disasters. At the meeting, it was established that, due to the principle of state sovereignty, "the activities of UNHCR will always be in support of the state, complementing rather than supplanting state responsibility" (Executive Committee of the High Commissioner's Programme 2011, p. 3). Also, the UNHCR's involvement is not exclusive, but happens within an inter-agency approach and on a mutual basis with other humanitarian agencies.

Another forum of inter-organizational cooperation is the Nansen Initiative, which also arose during the early 2010s. As one expert explains: "the Nansen Initiative was created because this issue [EDPs] could not be put under the scope and mandate of the work of the UNHCR" (SR1, para. 47). Those states which play a part in the debates on environmental displacement attach great importance to IOs participating in the global forums. Here, the additional benefit of IOs is to legitimize the state-led process of the Nansen Initiative and to give it some organizational support (SR1, para. 34–35). Additionally, it is perceived that the various IOs complement each other as regards content (SR4, para. 21–22). The UNHCR began to work closely together with the IOM, the IFRC, the NRC, with Walter Kälin in his role as Envoy of the Chairmanship of the Nansen Initiative from 2011 onwards, and with the UNU (IOR1, para. 7). Among others, the inter-organizational cooperation manifested itself in the joint policy brief of the UNU and the Nansen Initiative of 2014 called 'Integrating Human Mobility Issues within National Adaptation Plans'. In this inter-organizational publication, the depiction of EDPs clearly has a focus on human security; the policy brief even lists the terminology of human security in the chapter 'Key terms and concepts'. According to the participating actors, human security in the context of climate change means that "the core of human lives includes the universal and culturally specific material and non-material elements necessary for individuals to act on behalf of their own interests and to live with dignity. There is robust evidence that poverty, many kinds of discrimination and extreme natural and technological disasters undermine human security and that it can be enhanced by actions to reduce vulnerability to climate change" (Warner et al. 2014, p. 15).

This inter-organizational development is a good indicator that the relationship between the UNHCR and the IOM is rather cooperative. Concerning IO cooperation, one expert explained that "basically, if we want to push migration onto the climate change agenda, we have to do this together" (IOR4, para. 52). The staff working at the UNHCR and the IOM stress that the cooperation in the advisory group has been successful due to the personal relations and direct contact between the staff (IOR5, para. 63–66), the overall aim of the IO cooperation being "to get the topic on the radar and to get it into the climate policies, that is all" (IOR4, para. 54). In March 2014, the UNHCR collaborated with the Brookings Institution and Georgetown University and organized a consultation on planned relocations, disasters, and climate change. The intention was to identify major lessons learned and to provide further guidance to the UNFCCC on climate-induced planned relocations; planned relocation, migration, and displacement being the three types of human mobility to be considered by climate change adaptation measures (Advisory Group on Climate Change and Human Mobility 2014, p. 5). This so-called San Remo Consultation led to a background document with a strong focus on planned relocation. This document shows that the human security impetus guided the authors because it has a strong human-centered perspective and looks at the matter of climate change and relocation from the individual's perspective (Ferris 2014, pp. 7–9).

In the prelude to the COP in Paris in 2015, the advisory group reached the agreement that "human mobility be addressed in all its forms (displacement, migration, planned relocation) and as a matter of adaptation in the main agreement and its potential decisions" (Advisory Group on Climate Change and Human Mobility 2015, p. 2). Once again, the human security focus becomes apparent: on its 'Climate Change and Human Mobility Solution Agenda', published before COP 21 in Paris, the advisory group made individuals, households, and communities central as the beneficiaries of the recommended resilience solutions (Advisory Group on Climate Change and Human Mobility 2015, p. 1). During the side event before the COP in Paris, the UNHCR once again focused on the topic of planned relocation of populations (Advisory Group on Climate Change and Human Mobility 2015). That said, the UNHCR seems to have found its scope of work among the other international actors. As the topic of planned relocation is close to the organization's original work as the leading UN refugee agency, no other agency disputes the UNHCR's leading position on the matter.

A side-effect of the strengthened inter-organizational cooperation was that states also started discussing the topic (IOR4, para. 12). This also shows in the follow-up to the Nansen Initiative, the Platform on Disaster Displacement (PDD). The international community—NGOs, IOs, and a small number of committed states—acknowledged that the initial process of the Nansen Initiative must be kept alive. That said, the influence of the organizations to push the topic forward shows in the establishment of Article 14f of the Cancun Declaration, appears in the Sendai Framework, and becomes apparent in the Paris Declaration. By means of these inter-organizational forms of cooperation, the same group of people meets again and again over the years. As they all share expertise on the topic of EDPs, this group of mixed IO staff is much closer topic-wise compared to their fellow staff members inside their respective IOs (IOR5, para. 99–100). When talking to these individuals, it shows that they all acknowledge the advantages of cooperation and they all follow the goal of pushing migration and displacement onto the climate change agenda, which would be hard for any one of the organizations to achieve on its own (IOR4, para. 52). As one expert explains, "even if the UNHCR had the mandate, if you put it under the UNHCR, it would come out like a small agenda item, not as a big whole process" (SR1, para. 50–51).

In other words, the staff from the different IOs were "carving up the territory" (IOR3, para. 25–26) to achieve the goal of getting the topic on the radar of states (IOR4, para. 54). Much of the bilateral cooperation among the IOs aims to get the issue onto states' programs and to have a larger impetus in the international—state-led—forums (NGOR1, para. 27). That said, the IOM and the UNHCR especially perceive that there is so much work to do on the topic that it will not be helpful if they fight against each other. As one IO representative explains, "as there is so much to do, no one steps on anyone's toes" (IOR3, para. 51–54). Many of the international actors working on the issue—e.g., academics and state representatives—perceive that it needs both organizations, firstly because the IOM is not that bound by the Syria crisis and, secondly, because the UNHCR's capacities are not sufficient on their own to manage the topic of EDPs (AR1, para. 39 and para. 44–45). As one expert sums up, "we need to keep them [UNHCR and IOM] separate and also bring them together in a meaningful way" (IOR3, para. 49–50).

**Organizational Competition**

Despite these examples of intra-organizational cooperation, the UNHCR and the IOM also competed over which IO was the most fitting candidate to approach the topic. The UNHCR's interest in EDPs needs to be interpreted keeping the complex inter-organizational relationships in mind. Up until the 2000s, the UNHCR had been the leading agency to address human mobility; however, with the occurrence of, among others, the IOM, the organization was confronted with growing institutional competition (Betts 2009, p. 11). Jeff Crisp, a former senior UNHCR staff member and acclaimed refugee scholar, describes the competitive organizational environment as a 'humanitarian marketplace' where all the IOs, NGOs, etc. seek to "enhance their visibility, their fundraising potential, and hence their operational presence and impact" (Crisp 2009, p. 76). As a result, states were able to carry out 'forum-shopping', and this led to the UNHCR taking up new issue areas (Betts 2009, p. 11). Following this line of thought, the organization's interest in EDPs reflects that the UNHCR effectively participates in this 'humanitarian marketplace' (Crisp 2009, p. 76). In this context, High Commissioner Guterres clarified the difference between the UNHCR and other IOs (in this case, the IOM) since the beginning of his reign. In his opening statement at the ExCom in 2007, Guterres stated that the "UNHCR is not a migration management agency and does not want to become one. But to be able to fulfil our mandate, we must recognize the mixed nature of many present-day population flows" (Guterres 2007, p. 2).

On the one hand, the inter-organizational competition reveals itself on the question of terminology. Even though the Platform on Disaster Displacement has resolved some of the debates since its establishment in 2016, the terminology had been unresolved due to differing organizational perspectives until then (IOR4, para. 54). On the topic of the UNHCR's role in the field of environmental displacement, debate has taken place on the question of which organization is able to staff the topic (IOR1, para. 20–21) and which organization annexes the topic (SR6, para. 15–16). Moreover, some differing perspectives are present among the experts on whether the IOM or the UNHCR is the forerunner on the topic and who founded the inter-organizational cooperation (IOR1, para. 21; SR5, para. 13–14). As an example, when talking to IO representatives, different staff from different IOs describe the Nansen Initiative as their 'baby' (IOR1, para. 21; IOR5, para. 63–66). Some players in the international arena reveal that there was competition between the IOM and the UNHCR on the matter of displacement, going so far as to proclaim that this competition overshadowed the issue (NGOR1, para. 13). As one expert exclaims, "these agencies, they live by their reputation" (IOR3, para. 49–50). In other words, even though the official version of the organizational relations is that they all cooperated in harmony, some states and academics give hints that there existed, and still exist, debates on the question of organizational responsibility (SR6, para. 15–16). These rivalries among IOs are accredited to personal goals, visibility, or money (SR5, para. 18). Also, the struggle for reputation was mentioned as one example of inter-organizational competition (IOR3, para. 49–50). This inter-organizational behavior was also shown in the context of the COP in Copenhagen in 2009: some of the organizations still perceive that had they not chipped in and played a role in the process, some other organization would have taken this role; and as the funding always goes to the active ones, the organizations tried to have their say on the matter (NGOR1, para. 5).

On the other hand, it is also relevant how many staff the two IOs assigned to the topic of EDPs. In 2008, the IOM had the section on the environment, with one senior staff member and additional personnel of one and a half positions; by 2015, the IOM had six staff members and junior professional officers working within the team. The personnel working on the issue of EDPs were funded by the regular IOM budget. In contrast, the UNHCR has an environment section, but the focus of this section lies on the impact of refugee camps on the environment. On the issue of EDPs, the UNHCR has only had one person working on the topic and this person has also represented the organization at the Nansen Initiative. This position at the UNHCR is funded by the European Commission, which means that the person working on this job is not an internal UNHCR staff member but is recruited externally. For the follow-up to the Nansen Initiative, the Platform on Disaster Displacement, both the IOM and the UNHCR have one position that is funded by Germany because Germany chaired the platform's steering group during the first 1.5 years—i.e., from July 2016 to December 2017. Also, both organizations are standing invitees to the platform's steering group.

States which cooperated with the IOs on the topic tended to keep up the inter-organizational competition: from the states' point of view, every IO contributes to the debate according to its specific background and adds expertise to the discussion (SR4, para. 17–18). According to this cooperation argument, some states insinuate that were the topic of EDPs attributed to the UNHCR, including an enlarged UNHCR mandate, EDPs would be just one of many UNHCR agenda items (SR1, para. 50–51). Compared to the IOM, it took much longer for the UNHCR to find its level on EDPs and it was more challenging to get the internal backing on the issue (NGOR2, para. 57). Nevertheless, the UNHCR qualifies for the protection of EDPs because it holds the UN mandate for protection issues and is the head of the cluster system (AR1, para. 39). Similarly, the opinions of state representatives and academics differ on the question of whether the topic of EDPs should be transferred solely to the IOM (AR1, para. 39; SR1, para. 54–55; SR4, para. 17–18). The critics argue that the IOM has no protection mandate like the UNHCR, even though protection issues are an issue in the context of environmental changes and disaster displacement (AR1, para. 39). To put it more bluntly, as long as the cluster system exists, including its protection focus, it will be almost impossible all of a sudden to assign the topic and its protection implications to the IOM (AR1, para. 39). Some experts eye the recent publication of the IOM's own protection policy critically and argue that as soon as the money ceases to flow, the organization will drop the subject (AR1, para. 39). Thus, the recent protection approach of the IOM does not have a long protection history. Furthermore, cooperation between states and the IOM had, until 2016, been hampered because the IOM was not part of the UN system (SR4, para. 19–20). On the question of allocation of its resources, the IOM will need more funding if it is given a bigger role in the matter of EDPs (NGOR1, para. 22–23). Still, many experts stress that when it comes to logistics, the IOM works better and has good contacts. Outside Geneva, on the ground and in the field offices, the system would quickly fall apart without the IOM (SR3, para. 8).

A case study report about the UNHCR's involvement in disaster situations published in 2013 detects another aspect of IO competition: according to the author, research on the UNHCR's disaster involvement led to the conclusion that the greatest challenge for the organization is "simply establishing UNHCR as a credible actor within the natural disaster response framework of institutional actors and processes. In an environment that heavily focuses on quickly saving lives, other operational partners have a tendency to brush protection inventions aside as 'non-essential' or not 'lifesaving'" (Entwisle 2013, p. 53). Nevertheless, the case studies ultimately showed that by contributing non-food items—tents, etc.—the UNHCR was able to build credibility as an operational partner.

#### 5.2.1.3 The Influence of Hazards

The empirical data show that the UNHCR's interest in the topic of EDPs coincides with a number of hazards that occurred in the early 2000s and which received a lot of international attention (AR1; IOR3, para. 17–18). This empirical observation resulted in a new assumption for this section on the organizational environment:

New assumption: The occurrence of hazards influences the approach to EDPs.

The 2004 tsunami in Southeast Asia, the big earthquake in Pakistan, and the hurricanes and earthquakes in Haiti from 2004 onward put the issue of EDPs on the UNHCR's agenda. From the early 2000s until 2008/2009, the UNHCR's approach developed in parallel with high impact hazards that led to the displacement of large numbers of people. One expert presumes that "maybe this [the UNHCR's EDP approach] goes back to the tsunami in December 2004 in Southeast Asia which triggered a very strong call that we need to understand what happens with displacement in disaster settings" (IOR3, para. 6). Several other experts also suppose a link between the UNHCR's involvement with EDPs and the occurrence of hazards in general and the tsunami in 2004 in particular (AR1, para. 2–3). The international interest in these hazards put many IOs and NGOs on the ground, even though they do not usually work in disaster settings (AR1, para. 2–3; IOR3, para. 8–10). Still, there are two factors that explain their involvement: firstly, the international pressure due to the magnitude of the disaster and, secondly, the organization's already existing expertise and infrastructure on the ground (AR1, para. 10–11; IOR3, para. 8–10). More specifically, the concepts developed up to the late 2000s were all disaster concepts and, among others, resulted in the 'IASC Operational Guidelines for Protection from Natural Disasters' (IOR3, para. 17). From this perspective, the UNHCR's disaster involvement in the early 2000s was no different from that of other humanitarian actors. It shows how the UNHCR's work depends on external factors and the empirical data illustrate how some hazards have triggered the UNHCR's efforts on EDPs.

Building upon this hazard focus, the UNHCR's approach to EDPs also correlates with the international rise of interest in the topic of climate change. From 2009 onwards, the UNHCR's organizational environment changed and politicians, the civil society, the media, and the civil servants in the international committees debated climate change in preparation for the COP in 2009 in Copenhagen (IOR3, para. 17–18). In other words, the focus shifted from a purely disaster-centered concept to a climate change-thinking (IOR3, para. 17). Therefore, the topic of EDPs was largely seen from a climate change perspective as the media and NGOs used the term 'climate change' on a large scale. As one expert explains, "toward Copenhagen it became seen as more likely a climate change issue" (IOR3, para. 17). This change in how the international community perceived EDPs had started with a hazard focus and developed into a climate change debate up until 2009.

The organization's depiction of EDPs also reflects this changing environment. Prior to the COP in 2009, the organizational environment—i.e., environmental and humanitarian IOs and NGOs—acknowledged that, from an advocacy point of view, climate change functions as a big trigger. The actors who advocated the issue before Copenhagen had to take into consideration that large-scale and internationally visible hazards only happen once or twice a year (IOR3, para. 17). That said, international attentiveness to matters of environmental change only peaks a few times per year, which is not enough to provoke rethinking on the global scale. From an advocacy point of view, the problem with extreme events like hazards is that "as a policy priority, it kind of fades away until the next one [hazard] hits" (IOR3, para. 18). Therefore, many humanitarian organizations did not differentiate between a flood and an earthquake for their advocacy, and they did not differentiate between a hazard and climate change. However, as international attention to short-term disasters quickly declines, the advocacy argument outweighs the protection argument and the climate change storyline is more helpful to generate international interest in the matter of EDPs (IOR3, para. 18). In other words, "from a protection point of view, you cannot differentiate an earthquake from a flooding and to me it doesn't really make sense to say this is climate change and this isn't. But from an advocacy point of view, we can say this is about climate change" (IOR3, para. 17). This agenda-setting of the UNHCR also shows in its ideational depiction of the issue. In its 'The State of the World's Refugees' of 2012, the UNHCR writes that "more people are already displaced annually by natural disasters than by conflict, and the long-term effects of climate change are expected to trigger large-scale population movements within and across borders" (UNHCR 2012, p. 2). This description is rather unspecific but still attracts attention due to its dramatic future predictions. Thus, the focus does not lie on the individuals who suffer, which would imply a human security perspective, but rather on the enormous future population movements as impacting on security.

Table 5.1 depicts the tabular summary of Sect. 5.2.1 and gives an overview of the influence of the UNHCR's organizational environment.

### *5.2.2 The Influence of States on the UNHCR's Decision-Making*

This section presents the results from the empirical analysis concerning the assumption that states influenced the UNHCR in its efforts toward EDPs (for a summary of this section see Table 5.2). The following paragraphs present the insights gained from the coding of 'states' and puts the results into context with the information gained from the primary and secondary data. The analysis is based on the following theory-led assumption, as developed in Chap. 4:

(3) The UNHCR's 'member states' have an influence on the organization's EDP approach.

#### 5.2.2.1 International Politics Tearing at the UNHCR's Mandate

Regardless of the difficult and often state-centric playing field, the UNHCR was quite effective when it came to socialization of the newly independent states in the 1960s and 1970s. In order to become part of the international community, the UNHCR lured the new states in Asia, Africa, and South America with material incentives and moral persuasion to join the refugee protection regime (Loescher and

**Table 5.2** The influence of states on the UNHCR's decision-making

| Ideational depiction<br>(broadly informed by security theory) | Organizational factors<br>(operationalized organizational theory) |
|---|---|
| **2005–2011** | |
| – Many states presume that environmental changes have an impact on national and international security | – Criticism from states on the UNHCR's EDP approach based on the organization's existing workload/undermining the refugee mandate/ mission creep<br>– Until **2009**: A number of states (e.g., China, India, and the US) were still in denial about climate change |
| – **Conclusion:** Challenge for the UNHCR's EDP approach: states are the major source of funding; national interests of states tie the UNHCR's hands | |
| **2011–2015** | |
| – Diverse depictions of EDPs: some states focus on the suffering individuals and others on the security implications of human mobility | – **2010**: COP in Cancun: states acknowledge that displacement and climate change is an issue; states start discussing the topic at the ExCom<br>– **Standing Committee meeting in June 2011**: UNHCR document declares that it will support states in their work and it will only act upon states' consent; UNHCR argues that the funding for its EDP approach will not be at the disadvantage of refugees<br>– States are not convinced and encourage the UNHCR to focus primarily on its mandated responsibility<br>– **ExCom meeting in Oct. 2011**: States are thrown off guard by the UNHCR's approach and are noncommittal on the topic of EDPs; reluctance to let an IO look into the issue; fear that this could result in an obligation to pay<br>– **Ministerial Meeting in Dec. 2011**: States influence the UNHCR's EDP approach; Guterres wants states to adopt a ministerial declaration<br>– States counter that the UNHCR only wants to 'refugize' the topic and wants to create a refugee problem to find new clients<br>– Group pledge by Switzerland, Norway, Costa Rica, Germany, and Mexico; Norway and Switzerland fund the staff position at the UNHCR<br>– **2012:** States could support the topic in the Nansen Initiative without losing face |
| – **Conclusion A**: UNHCR's difficulties with funding: to balance the wishes of the donor states of the global North vs. the refugee hosting countries of the global South<br>– **Conclusion B**: The US is the largest financial contributor and the hegemon within the global refugee regime<br>– **Conclusion C**: States did not have a stake in the preliminary steps of the organization toward EDPs; the Nansen Initiative states play a role in the UNHCR's intensifying EDP approach | |

Milner 2011, p. 195). Until the 1980s, the UNHCR "enjoyed maximum legitimacy as it simultaneously tried to define the refugee issue for states, to convince governments that refugee problems were soluble, to prescribe solutions, and to monitor their implementation" (Loescher and Milner 2011, p. 194). The work of the UNHCR during the 1980s centered on protracted refugee situations in Asia, Africa, and South America. This shift expanded the original focus of legal protection toward refugee camps.

In the 1990s, the organization had to tackle the problems that resulted from the demise of the Soviet Union and the UNHCR involved itself in cases of repatriation and humanitarian relief. The various crises of the 1990s—e.g., in the former Yugoslavia and Somalia—revealed how state-led refugee protection—or rather, the lack thereof—reached its limits and led to a rise in requests for the UNHCR's involvement (Loescher and Milner 2011, p. 196). Since the 1990s, the policies from the global North have challenged the organization's work as they aim to restrict the influx of refugees and try to contain refugees in the global South. Examples of the restrictive policies of the global North are "non-arrival policies, such as safe third-country agreements, an increasingly restrictive application of the 1951 Convention, and a range of deterrent policies, such as detention of asylum seekers and the denial of social assistance" (Loescher and Milner 2011, p. 197). In addition, the reactions in the refugee host countries of the global South complicate the work of the UNHCR. In the 1990s, the countries of the global South tried to limit the number of refugees on their territory because they experienced very large numbers of people in need of refuge and these people tended to stay in the host countries for longer periods of time. In this context, it is important to factor the changing composition of the ExCom into the analysis: whereas the developed states outnumbered the committee for a long time, the composition changed in favor of the developing states; this also led to the countries doing the initial take-up of refugees being better presented in the ExCom (Dingwerth et al. 2014, p. 195). Therefore, the national sentiments of both the global North and the global South largely influenced the work of the UNHCR and triggered how the organization interpreted its mandate.

Throughout the period from 2005 to 2015, many states acknowledged that environmental changes have an impact on national and international security. Despite this acknowledgement and as regards the changing interpretations of the UNHCR's mandate, the debate on questions of mandate adaptation and expansion was, and still is, immense. The debate ranges from some states that question the organization's moral authority on matters of refugee protection to other states that only pursue their national interests when it comes to questions of asylum and refugee policies (Loescher and Milner 2011, p. 194). This debate also resonates in the context of EDPs: here, the critical remarks of states boil down to the argument that "the UNHCR is already too busy with the refugee crises" (IOR5, para. 19). According to one IO representative, this critique faded away during the last years and states apparently now understand why the UNHCR involves itself in the topic of EDPs (IOR5, para. 19).

#### 5.2.2.2 National Politics Tearing at the UNHCR's Mandate

The struggle that the UNHCR has faced since its foundation is to strike a balance between its highly political role of supervising states in their fulfillment of the Refugee Convention and fulfilling its own standard of being an apolitical humanitarian organization (Glover and Russell 2001, p. 5). As Glover and Russell argue, "performing one function well practically precludes UNHCR from performing the other function as well" (Glover and Russell 2001, p. 5). Furthermore, these two functions have to be fulfilled against the background of states being the major source of funding for the organization. States are free to contribute to the annual budget or to ad hoc special appeals, whatever might be closer to their respective national interest. As Betts puts it, "without any binding obligation to contribute to burden-sharing, states have selected their priority areas and earmarked their contributions to UNHCR projects and programs in accordance with their own perceived interests" (Betts 2009, p. 16). To put it differently, the political realities and national interests of states tied—and still tie—the UNHCR's hands; the new issues that the organization has tackled throughout the years thus mirror dominant state behavior. Moreover, because the funding by states is voluntary, the UNHCR has been dependent upon a small number of states from the global North which have paid the largest proportion of the organization's expenses (Loescher and Milner 2011, p. 196).

States have not always welcomed the expansion of the UNHCR's work into the issue area of EDPs, the underlying arguments being, first, that states could exploit the UNHCR to react to natural disaster situations, even though states have the responsibility and the obligation to protect their citizens from rights violations after disasters and the obligation to mitigate the risks of natural disasters, as determined in the Hyogo Framework of 2005 (Entwisle 2013, p. 12). Second, states were critical about the UNHCR's opening to EDPs because the organization's commitment to a variety of causes of flight could potentially undermine its original refugee mandate (Loescher and Milner 2011, p. 194). Moreover, some states and key stakeholders also have concerns regarding anything that looks like mission creep (Goodwin-Gill and McAdam 2017, p. 6). On the other hand, those states which already suffer under the impact of environmental changes, like Nauru, prominently called for "all the organs of the United Nations to work collectively within their mandates and to intensify their efforts" (UN General Assembly 2009a, p. 3) in the UN GA in 2009. Third, states had already been critical about the issue of IDPs because they feared an intrusion into their sovereignty (AR1, para. 20) and they did not like to hear that they had IDPs displaced by conflict; instead, the governments argued that the people were displaced by droughts (IOR2, para. 26). Even though the UNHCR ultimately decided to include conflict-IDPs as persons of concern, the UNHCR staff fiercely debated the question of including disaster-IDPs (AR1, para. 15).

**Detailed Chronology of States' Influence**

The Nansen Initiative conference took place in June 2011. The underlying intention of the states and IOs that involved themselves in the initiative was to "have more knowledge about the issue and to be able to conceptualize it and identify possible solutions" (IOR5, para. 10). Every participating state had a different affinity with the IOM or the UNHCR, these two IOs being the biggest to sit in the steering group (SR5, para. 18). From the perspective of one IO representative, two factors pushed the issue of EDPs at the global level: firstly, the states which picked up the topic after the ExCom meeting and started the state-led process in the Nansen Initiative; and, secondly, the IO staff who worked on the ground and who saw disaster and conflict displacement intermingling on a daily basis (IOR5, para. 27). With the establishment of the Nansen Initiative, more and more states were able to support the topic without the loss of face of a UNHCR mandate-clashing initiative (SR1, para. 15). It was crucial to a lot of states that the process was state-led (SR6, para. 14).

In June 2011, the UNHCR prepared a document called the 'UNHCR's Role in Support of an Enhanced Humanitarian Response for the Protection of Persons affected by Natural Disasters' (Executive Committee of the High Commissioner's Programme 2011) to present to the Standing Committee and the ExCom meeting the same year. The paper provided a policy framework in which the UNHCR outlines its engagement in natural disasters. This includes a set of general considerations, the first two of which concern the UNHCR's relationship with the respective state in which the natural disaster situation occurs. Here, the UNHCR makes clear that "at all times, the responsibility of the State for citizens and others on its territory remains primary. UNHCR's engagement thus serves to support the role and responsibility of the State" (Executive Committee of the High Commissioner's Programme 2011, p. 6). In addition, the "UNHCR's engagement—its commencement, its continuation and its geographical scope—is premised upon State consent" (Executive Committee of the High Commissioner's Programme 2011, p. 6). The UNHCR mentioned these two guiding considerations of its engagement in natural disasters at the beginning of the publication to make clear that it was well aware of possible state sensitivities when it comes to their sovereignty. Also, the UNHCR anticipated the states' questions about funding and therefore included the following consideration in the paper: "Funding for the response to natural disasters will come from inter-agency flash appeals, separate UNHCR appeals, and the Central Emergency Response Fund. Thus, the extent of UNHCR's operational protection response may be limited by funds raised directly in relation to the disaster" (Executive Committee of the High Commissioner's Programme 2011, p. 7).

The reactions of states during the meeting of the Standing Committee in June 2011 illustrate that the majority of committee members questioned the pilot arrangement of the IASC that designated to the UNHCR the role of lead agency for coordination of the protection response in situations of natural disaster (Executive Committee of the High Commissioner's Programme 2011, p. 7). As the draft report about this meeting reads, almost half of the delegates declared their "support in principle (albeit with many reservations related to the desirability or not of UNHCR taking on such additional responsibilities and activities)" (Executive Committee of the High Commissioner's Programme 2011, p. 7). Furthermore, "the majority of Committee members declared that they were not convinced by the information provided to date. Several delegates encouraged UNHCR to focus primarily on its mandated responsibilities. Others called for further independent evaluation, as well as discussion amongst agencies and States [...]. There was a clear call for postponement of any designation of responsibility [...]" (Executive Committee of the High Commissioner's Programme 2011, p. 7). The draft report therefore gives an account of the concerns mentioned by the delegates: state sovereignty, resources and capacity, mandate implications, and points of disengagement. Responding to states' remarks, the Assistant High Commissioner emphasized that further consultations would be convened on the question and that the UNHCR would prepare a paper about why the IASC proposed this new role for the organization and why it would be good to move away from the current system of ad hoc arrangements of the cluster approach.

The topic was not picked up at the follow-up meeting (52nd) of the Standing Committee in September 2011 and natural disasters were not mentioned (Executive Committee of the High Commissioner's Programme 2011). The draft report of the meeting of the Standing Committee in September 2011 reveals that "with regard to the Office's role in natural disasters, several delegates cautioned that this should not be at the expense of its core mandate; that it should be subject to the consent of the Government concerned; and that it should not entail any diversion of funding from core mandate activities. UNHCR was urged to work on the question in close coordination with other agencies and to brief the Standing Committee" (Executive Committee of the High Commissioner's Programme 2011, p. 5). This response to the Director of the Regional Bureau for Asia and the Pacific of the UNHCR shows the above-mentioned reluctance of states and led to the Director assuring the respective states and delegates that the UNHCR would only act in an inter-agency context on the issue of natural disasters. He also agreed with the delegates' remarks that even though the UNHCR was the global lead in the cluster approach, the organization was not automatically engaged in the matter of disasters. Lastly, the Director responded that the UNHCR's role in some situations would only be advisory in nature and would always depend on the respective government's response plan (Executive Committee of the High Commissioner's Programme 2011, p. 5).

The ExCom meeting in October 2011 made it clear that, overall, the states were thrown off guard by the UNHCR's recommendations and were rather noncommittal to the topic of EDPs (IOR5, para. 10). In this context, the argument surrounding the UNHCR's mandate was repeatedly made by state representatives. According to many states, the UNHCR had a clear mandate that was written in the 1951 Convention. Following the Refugee Convention, "the situation of environmentally displaced or climate related displacement or migration or whatever you call it, does not fall into that category of persecution that makes them refugees" (SR1, para. 15). At that time, states were concerned about questions of attribution and compensation if they pushed the topic of EDPs. As a consequence, they were reluctant to let an IO look into the issue if this could actually result in states being pressured to sign up to or create an obligation to pay (NGOR2, para. 35). That said, some of the developed states feared that they would have to pay for people in Bangladesh who had been displaced or that they would have to compensate small island states (NGOR2, para. 35). During an international conference in 2013, José Riera of the UNHCR commented on the 2011 Ministerial Meeting and said that "the majority of states clearly signaled that they were not ready to discuss, elaborate or accept new international obligations in this regard" (Riera 2013a, p. 15). Herein shows the disappointment of the UNHCR staff who were working on the topic.

In the report on the ExCom meeting of October 2011, the topic of EDPs is not even picked up—only the Chairman's summary of the debate in the annex of the ExCom's report about the meeting reads that the "member states also recognized that UNHCR works in an increasingly complex environment. Trends such as climate change, natural disasters, food and water insecurity, as well as growing vulnerabilities and armed conflicts, stretch the humanitarian system" (Executive Committee of the High Commissioner's Programme 2011, 8). Still, states' contributions to the ExCom meeting in October 2011 revealed that a lot of states did indeed acknowledge that there was a protection leadership gap and that the UNHCR was an important actor in this respect (Goodwin-Gill and McAdam 2017, p. 19). Many delegates' remarks referred to the topic of climate change and disasters—among others, Spain and Germany—and to the topic of environmental displacement—among others, Norway and Mexico (Goodwin-Gill and McAdam 2017, p. 19). Contrary to this, the Russian Federation and India cautioned about possible changes to the UNHCR's mandate and emphasized that refugee/asylum and migration issues should remain separate (Goodwin-Gill and McAdam 2017, p. 20).

Another example which exemplifies states' influence on the UNHCR's EDP approach is the UNHCR Ministerial Meeting to commemorate the 60th Anniversary of the Refugee Convention in December 2011. During this meeting, with all its high-level participants and state representatives at one table, the UNHCR brought forward the issue of EDPs (IOR5, para. 10). In his opening statement, High Commissioner Guterres explicitly mentioned that the effects of climate change—e.g., natural disasters and desertification—uproot and displace a growing number of people. Guterres therefore sees "a number of serious protection gaps, particularly in the context of [...] environmentally-related displacement" (Guterres 2012, p. 141). Referring to the Nansen Conference a few months prior to the UNHCR Ministerial Meeting and to the Nansen principles, Guterres declares that "we are ready to work with interested states and other relevant actors to help develop such guiding frameworks" (Guterres 2012, p. 143). The UNHCR intended to use the celebration of the commemoration of the 60th anniversary of the Refugee Convention to make states adopt a ministerial declaration or a ministerial communiqué that acknowledged that environmental changes were a new sort of driver of human mobility and movement. However, as one IO representative remembers, both the developed countries—i.e., the donors—and the countries from the global South told the IO staff to "jump into the lake" (IOR1, para. 9). In this context, states were reluctant to adopt a ministerial declaration on EDPs because they possibly "did not want to create a refugee problem where there was none" (IOR1, para. 9) and "they suspected that the UNHCR was trying to 'refugize' this whole area" (IOR1, para. 9). In other words, states were afraid that the organization was "trying to find new clients for the organization" (IOR1, para. 9). According to this interpretation, states perceived that "the UNHCR was already at pains to meet the needs of the refugees" (IO1, para. 14). The perception of some of the states was that the UNHCR was already hardly able to deal with the refugees and therefore these states did not want the organization to move into other problem areas (IOR3, para. 22). Consequently, states called on the UNHCR to "focus on your core business and do a good job at that before you are looking left or right" (IOR1, para. 14). Adding to this skepticism on the workload of the UNHCR, a group of states was not happy about the UNHCR being too prominent. One state representative describes this skepticism as follows: "They are afraid to lose control—and sovereignty is very important to them [states]" (SR6, para. 19).

However, even though then High Commissioner Guterres speaks of a "quantum leap forward" (UNHCR 2012, p. 6) in relation to the protection of stateless people, the states at large responded rather critically to the topic of EDPs in 2011. This shows, amongst others, in the Ministerial Communiqué adopted at the meeting in December 2011 which does not specifically pick up the topic of EDPs (Kälin and Schrepfer 2012, p. 55; Ministerial Communiqué 2011). The underlying reason is that "most of the states were a bit reluctant" (IOR5, para. 10) and asked themselves "what is this issue and what does it have to do with the UNHCR, which already has a lot to do with conflict displacement, and we don't know anything about climate change disaster displacement, how big is that?" (IOR5, para. 10). Critical responses came from both the refugee host countries and donor countries. As an example of the former, Bangladesh feared that if the UNHCR were to open its mandate to other triggers of flight, the UNHCR's support of Bangladesh would diminish or disappear.

Nevertheless, the pledges made by states before, during, or after the Ministerial Meeting in December 2011 also include the group pledge by Switzerland, Norway, Costa Rica, Germany, and Mexico on the topic of "people displaced across borders owing to sudden-onset disasters linked to climate change" (UNHCR 2012, p. 38). These countries pledged "to cooperate with interested states, UNHCR and other relevant actors with the aim of obtaining a better understanding of such cross-border movements at relevant regional and sub-regional levels, identifying best practices and developing consensus on how best to assist and protect the affected people" (UNHCR 2012, p. 38). This small group of states wanted to approach the question: what do we do with people displaced due to natural disasters and climate change? Their interest ultimately resulted in the establishment of the Nansen Initiative. During the Ministerial Meeting, the President of the Swiss Confederation, Micheline Calmy-Rey, delivered a statement in which she summed up the intention of Switzerland and Norway to better understand external displacements resulting from natural disasters (Calmy-Rey 2012, pp. 147–148). Similarly, Norway repeated the Norwegian-Swiss pledge and called for better international consistency and cooperation to protect people displaced externally due to sudden-onset disasters (UNHCR 2012, p. 102). Latvia's pledge included the belief that the global consequences of climate change should be given serious consideration (UNHCR 2012, p. 89) and Germany "welcomes UNHCR's recognition of climate change as an additional trigger of displacement" (UNHCR 2012, p. 76). Lastly, Mexico used almost identical wording to Norway and pledged to cooperate with the above-mentioned states to identify best practices and to develop a consensus on how to assist EDPs (UNHCR 2012, p. 96). Even though the participating states did not find a common understanding concerning how to approach EDPs—and thereby disappointed the UNHCR's prior expectations about the meeting—Guterres nevertheless picked up the topic in his closing remarks by using the words: "we are witnessing an enormous increase in the number of people displaced by conflict, natural disasters and other environmental hazards" (Guterres 2012, p. 161).

As regards the state-IO cooperation in the Nansen Initiative, the leading states knew that they had to work strategically and to keep it as a state-led process because otherwise a lot of states—especially from the global South—would leave the Nansen Initiative and stop cooperating with it (SR3, para. 39). Some state representatives even speak of a dilemma with regard to the Nansen Initiative and its relationship with the UNHCR and with the IOM (SR1, para. 14). The strong role of states in this context shows in the 'Nansen Conference Summary', which clearly states that "it was suggested that States, in conjunction with UNHCR and other relevant stakeholders, could develop a guiding framework or instrument for the protection of people displaced externally due to sudden-onset natural disasters, including those related to climate change" (Nansen Conference 2011, p. 4). Adding to the importance of state leadership, the agreed-upon Nansen principles clarify that states carry the primary duty for the protection of their population and for those affected by environmental changes (Nansen Conference 2011, p. 5). Some IO representatives perceive that, after 2015, more and more states changed their opinion and are nowadays supportive of the Platform on Disaster Displacement (IOR5, para. 23). As one state representative sums up: "So for us, be it refugees, be it migrants, whatever ... the main concern is to look at the people and how we can give them some response. We cannot just ask them to go perish in the sea. So, this is the main concern and this is how we look at all these initiatives, the Nansen Initiative, etc." (SR1, para. 15). As another state representative explains, the UNHCR's mandate should be discussed and applied in a flexible manner, the aim being to interpret the mandate in order to fit the mandate to the actual situation at hand (SR1, para. 23). In other words, sometimes it needs "a different perspective to implement your mandate, not necessarily a different mandate" (SR1, para. 31). This change in states' opinions becomes clear when looking at Bangladesh: as of 2015, the country is, together with Germany, leading the follow-up organization of the Nansen Initiative, the Platform on Disaster Displacement.

#### 5.2.2.3 Funding Issues

The problem surrounding the UNHCR's EDP expansion goes back to the difficult position of the UNHCR in balancing the wishes of the big financial contributors of the global North on the one hand with the specific need to help the refugee host countries of the global South on the other (Loescher and Milner 2011, p. 196). Some states feared that the UNHCR would abandon them and would leave those states alone in their dealings with refugees and IDPs (IOR1, para. 14; SR3, para. 27). Additionally, the example of the US shows that, because this country had been the biggest donor since 1991 (Hüfner 2015, pp. 113–114), the US had a powerful position in the UNHCR. For decades, the US had contributed more than 30% of the organization's budget; consequently, the country acted as the "hegemon within the global refugee regime" (Betts et al. 2012, p. 97) and thereby had considerable influence on the UNHCR's activities and personnel decisions. The US was critical because it already contributed a lot to the UNHCR's budget and was afraid that the costs would rocket upward if the UNHCR put more on its plate (IOR5, para. 23; NGOR2, para. 33). As shown in Sect. 5.2.3, throughout the reign of Guterres, the position of the Deputy High Commissioner was continuously staffed by American citizens. Although this does not automatically imply that the UNHCR staff only followed their home country's policies, it is still relevant to incorporate the UNHCR's staffing decisions into the analysis. The perception of IO and NGO representatives is that, in 2011, states feared that the UNHCR was already overstretched; "and if one of your biggest donors says no, this is not what you are going to do" (NGOR2, para. 33).

The monetary argument also shows in the context of EDPs. As one IO representative explains, "the concern [of states] was not about 'should we deal with that?' but it was more a concern of 'do we have the capacity?'" (IOR5, para. 19). That said, the reluctance of many states concerning the involvement of the UNHCR in the matter of EDPs arose from funding issues. At that time—in 2011—the states expected that just one agency would approach the issue and so many states asked, "how will the UNHCR continue their mandate if they deal with that [EDPs]?" (IOR5, para. 33). During the meeting of the Standing Committee in 2011, "several delegates requested further information on how the UNHCR envisaged managing new areas of protection needs and related financial implications, notably [...] for pilot programmes, such as those for persons displaced as a result of climate change and natural disasters" (Executive Committee of the High Commissioner's Programme 2011, p. 14). Throughout the Nansen Initiative process, the EDP approach evolved into a multi-stakeholder topic and states came to an understanding that all agencies had to address the issue in part (IOR5, para. 33). This agreement took the pressure off the UNHCR. Therefore, as one state representative sums up, the states guaranteed sufficient organizational support to the IOs: the Nansen Initiative states created the Steering Group, nominated the envoy (Walter Kälin), and gave a policy background to the existing organizational approaches on the topic (SR6, para. 25). The intention was to have the IOs as standing invitees and not as members, because a number of states "were always reticent when an IO is trying to push a subject" (SR6, para. 25). As a consequence, the states got a guarantee that they could check the final document before publication, which made them powerful stakeholders in the EDP process (SR5, para. 10). States also funded the UNHCR's staff position on the topic of EDPs—the expert working on matters of EDP at the UNHCR was not funded by the UNHCR; rather, from 2011 onwards, Norway and Switzerland funded the staff and Germany took over the funding from 2016 onwards after the end of the Nansen Initiative. The intention was to institutionalize the expert position at the UNHCR until 2018 (IOR5, para. 85). To sum up the UNHCR's role at the beginning of the Nansen initiative: "Their [the IOs'] advice was very important ... but they had to have a concerted status instead of being really part of the group" (SR6, para. 27).

#### 5.2.2.4 States' Perspectives on EDPs

Another reason for states' reluctance for the UNHCR's EDP involvement was that they did not know a lot about the issue. In 2008, countries such as China, India, and the US were still in denial about climate change (IOR1, para. 20). As one IO representative explains, the opinion of states was that "we don't know about the issue, it is too early to talk about it" (IOR5, para. 33). From the perspective of one IO representative, the relationship between states and IOs developed as follows. IOs such as the IOM and the UNHCR first tried to put the topic of environmental migration and displacement on states' agendas; following this preliminary step, states nowadays ask these IOs about the latest publications and for state of the art information (IOR4, para. 8). From the states' perspective, the state-led process on EDPs incorporated the knowledge gained in the global regions of South America, East Africa, East Asia, the Pacific, South Asia, and Europe (SR5, para. 8). Due to the input from various states, the civil society, academics, regional organizations, and IOs, the Nansen process has been repeatedly called a bottom-up approach by states. Nevertheless, some state representatives say that it was more important that states approved because they were the ones who funded the initiative (SR5, para. 17). Compared to states, the IOM and the UNHCR should also identify with the concepts in order to commit to the initiative and support states (SR5, para. 17). Some states even describe their own role as being similar to a puppet master of the IOs, because these states perceive that the role of the IOM and the UNHCR in the Nansen Initiative was to mainstream the topic into their respective IOs (SR3, para. 31). Therefore, states also acknowledged the work of the IOs—for example, that they succeeded in establishing the topic in the Sendai Framework and in the Paris Agreement. Some states intended to push the IOs on how they approached the topic of EDPs (SR5, para. 24). In this context, one state representative claims that the states expected more of a performance from the UNHCR (SR3, para. 31): this statement is controversial as it very much contradicts the UNHCR's own perspective on its performance concerning EDPs. The relationship between the UNHCR and the states also has to do with trust: "We gained trust because we built trust for all these years. And also, because we achieved what we wanted, having the hook in the international agreements. [. . .]. So, having displacement in one of the big agreements will automatically bring states to have some interest in that issue" (IOR5, para. 67).

As one way to counter the pull of national interests, the UNHCR increasingly cooperated with other global actors. One example is the involvement of non-state actors in the UNHCR's field of expertise, which explains the adaptation of the UNHCR's field of work up to the mid-2000s (Loescher and Milner 2011, p. 194). The cooperation of IOs with humanitarian NGOs on issues of refugee protection is an example of how the UNHCR was—and still is—able to counter the interests of the state. One example of the successful cooperation of the UNHCR with other relevant actors is paragraph 14(f) in the Cancun Declaration. According to one IO staff member, the establishment of this paragraph in the global agreement changed how states thought of the issue. In other words, "thanks to that, because we had 14(f), states started to think, displacement and climate change, okay, that is becoming an issue. But we do not know anything about it. And that is how they [states] also started discussing at the ExCom" (IOR5, para. 71).

Looking at the timeframe between 2011 and the end of the Nansen Initiative in 2015, it is seen that states acknowledged that environmental movements were already recognizable and about a quarter of the states, which were UN members, had to deal with movements linked to disasters (IOR1, para. 17). According to state representatives, the UNHCR willingly agreed to be on the backseat, as one state representative said, "for the benefit of the people" (SR6, para. 43). By implementing the Protection Agenda, states gave the UNHCR and the IOM a reference to refer to in concrete situations and for their activities on the ground (SR6, para. 31). In addition, the leading states on the topic justified their powerful position compared to IOs by arguing that the Protection Agenda would not implement itself (SR2, para. 51).

Still, a lot of states were reluctant to approach the issue of environmental migration and displacement, even during the lifespan of the Nansen Initiative. This reluctance has to be interpreted against the background of the topic of migration not being part of the UN system until 2016. The question of to what extent the UN should tackle migration issues was still open for discussion and a highly sensitive topic (AR1, para. 56). The reason why the Nansen Initiative and the Global Forum on Migration and Development were situated outside the UN family for such a long time goes back to states' reluctance concerning their migration policies (AR1, para. 56). By keeping migration out of the UN system, states kept their sovereignty on the issue and thus did not push for a universal migration regime (AR1, para. 56). Compared to this, the refugee-producing countries and the states with a humanitarian tradition intended to push the migration issue onto the UN (AR1, para. 58). As one IO representative sums up: "We needed to be very careful because states showed negative reactions and reluctance actually for most of the time that the Nansen Initiative was running" (IOR1, para. 20).

Table 5.2 depicts the tabular summary of Sect. 5.2.2 and gives an overview of the influence of states on the UNHCR's decision-making.

### *5.2.3 The Character of the UNHCR's Bureaucracy*

The state-centric interpretation outlined above holds some deficiencies because it denies the UNHCR any kind of say in its work. Therefore, it becomes paramount to learn from existing organizational research that sheds light on both the individuals inside IOs and the organizational agenda-setting that the leadership and the staff get to do.

The UNHCR launched the Global Consultations on International Protection to discuss the protection issues that arose during the 1990s together with states, NGO representatives, and refugee experts. The outcome of this process is the Agenda for Protection, which specifies the necessary actions by states, NGOs, and the UNHCR to tackle the challenges of refugee protection in the twenty-first century. However, due to internal criticism of the Global Consultations—many UNHCR staff perceived the Agenda as rather superficial, without a focus on the individual, and as a prestige project of High Commissioner Ogata and the protection section at the UNHCR—the organization itself did not fully commit to the outcome (Loescher and Milner 2011, p. 198). The internal struggles concerning the interpretation of the UNHCR's mandate manifested themselves in the Convention Plus Initiative of 2001: newly elected High Commissioner Lubbers started this initiative even though the Global Consultations process was still ongoing. Donor states criticized this step by the High Commissioner because they had promised to financially support the unfinished Global Consultations (Loescher and Milner 2011, p. 199). As a result, opinions differ as to whether the intention of the Global Consultations and the Agenda for Protection to find collaborative solutions for refugee crises was a success or a setback. Irrespective of how one assesses these initiatives of the early 2000s, donor states' control of the organization's budget and the associated influence on the UNHCR's overall direction played a huge role. Against this background, it is important to look into the internal workings and the bureaucracy of the UNHCR. This section focuses on the following assumptions concerning the UNHCR's approach toward EDPs (for a summary of this section see Table 5.3):

(4) The UNHCR approaches the topic because it overlaps with the organization's goals.
(5) The UNHCR's headquarters push the organization's EDP approach.
(6) Individuals in the organization push the UNHCR's EDP approach.
(7) The High Commissioner pushes the approach and has a personal motive for doing so.

These assumptions will be approached by analyzing the following coding: 'organizational goals', 'headquarters influence', 'individuals', 'staff learning', and 'the High Commissioner'.

#### 5.2.3.1 The UNHCR's Organizational Goals

(4) The UNHCR approaches the topic because it overlaps with the organization's goals.

According to state representatives, the UNHCR pursued the organizational goal to "proliferate their [UNHCR's] mandate and grab some more work" (SR1, para. 22). The perception among the experts is that the UNHCR's organizational goal was to develop itself into a general forced displacement organization (AR1, para. 35), the

**Table 5.3** The character of the UNHCR's bureaucracy

| Ideational depiction (broadly informed by security theory) | Organizational factors (operationalized organizational theory) |
|---|---|
| **2005–2011** | |
| – **2007 and 2008**: Guterres' ExCom statement, 'Foreign Affairs' article and speeches portray a securitizing tendency of EDPs and draw on military jargon; dramatic description of the future number of EDPs<br>– **2008**: Deputy HC Johnstone bluntly declares in which cases the UNHCR has the responsibility to protect; he argues on the grounds of humanitarian suffering; dramatic and alarmist description of millions of displaced people | – The US staffs the position of Deputy HC until 2015<br>– UNHCR was called to help in natural disasters in 2004, 2006, and 2008 (external push)<br>– The UNHCR's opening toward IDPs was a prelude to the organization's opening toward EDPs<br>– Lack of thematic overlap and collaboration between the different units at the UNHCR; staff do not actively back up Guterres' task force on climate change<br>– Internal struggles due to top-down decision-making of Guterres concerning EDPs |
| – **Conclusion A**: UNHCR staff were skeptical and reluctant about the EDP topic (beyond the UNHCR's mandate)<br>– **Conclusion B**: High Commissioner Guterres picked up the topic during his reign; the rise of interest in EDPs at the UNHCR goes back to Guterres | |
| **2011–2015** | |
| – **2011**: Guterres' speech at the Nansen Conference depicts international security references but also refers to human security; he links EDPs with refugees and thereby draws attention to the UNHCR in the run-up to the Nansen Initiative<br>– **2011**: Guterres links climate change and displacement with international peace and security in his speech at the UN SC<br>– **2013**: José Riera is against the "parade of horrors" argument when it comes to EDPs; he speaks out for a human security depiction<br>– **2014**: Volker Türk extensively uses human security references in his public statements on behalf of the UNHCR | – Geneva headquarters are influential and try to raise the topic of EDPs with states and colleagues; some field offices also have the topic on their radar<br>– No institutionalized linkage of the topic between headquarters and field offices (top-down approach); ad hoc briefings<br>– **2009–2011**: Another peak level of UNHCR involvement in refugee flows and IDPs<br>– **Post-2010**: Guterres tried to institutionalize the topic at the UNHCR (task forces, publications, etc.)<br>– Guterres tries to push the relevance of the UNHCR by picking up the topic that he perceived as central in the twenty-first century<br>– Since **2011**: More balanced and fact-based publications by UNHCR staff (no more securitizing tendencies of EDPs)<br>– Until **2015**: Some staff learning process; the UNHCR commits to approaches through the COP process and the Nansen Initiative<br>– No general staff learning process: from 2009 onward, merely personal initiatives, briefings, and newsletters initiated by some committed staff inside the UNHCR<br>– **2015**: More awareness of UNHCR staff that the topic is interlinked with the UNHCR's mandate |

(continued)

**Table 5.3** (continued)

| Ideational depiction (broadly informed by security theory) | Organizational factors (operationalized organizational theory) |
|---|---|
| – **Conclusion A**: Even though only a small group of people worked on the topic at the UNHCR, the group had good access to the High Commissioner<br>– **Conclusion B**: It largely depended on the committed individuals at the UNHCR that the IO could establish itself so prominently internationally; but the engagement of some staff members was not reflected in the whole organization: the topic did not become institutionalized until 2015 | |

underlying reason being that this was where the organization's expertise lay and because there was a decline in the number of refugees as compared to the 1990s. Being aware of this critique and allegation, High Commissioner Guterres stated in his 2008 publication that the "UNHCR is not seeking an extension of its mandate, but it is our duty to alert the international community to the protection gaps that are emerging" (Guterres 2008, p. 9). Prior to Guterres' statement to the ExCom in 2009, it seemed that "there has been practically no institutional 'appetite' for involvement in natural disasters. The international community's focus on climate change and natural disasters has changed that, and, it would appear, so have institutional considerations about UNHCR's future role as the protection agency" (Deschamp et al. 2010, p. 3). In this regard, the organization's relevance was one aspect that led to the UNHCR's approach toward EDPs (AR1, para. 26). That said, the principles of the different IOs who came together in the IASC acknowledged that due to the large-scale disasters and the prospective climatic changes, something in their respective organizational environment had to change. Together with the other IOs, the UNHCR's goals thus were, first, to push states to act more responsibly in matters of displacement due to all kinds of triggers and, second, to realize more funding, because the traditional mechanisms were not enough (AR1, para. 31). Combined, these two goals resulted in extensive lobbying to the climate secretariat to take up the topic. Here, it can be noted that the argument that a monetary consideration lay behind the UNHCR's extension of its 'population' was not new: the report on the 1998 Conference concerning the UNHCR extending its mandate reads that some IO and NGO conference participants "felt that there are very powerful trends within UNHCR to maximize its budget and power through an assertion of its role way beyond its mandate" (Working Group on International Refugee Policy 1999, p. 413).

The UNHCR staff itself paint a different, more people-centered picture of its EDP approach. According to the UNHCR staff, it is not enough to envision a lone white polar bear when thinking about climate change; instead, it is largely about human beings (IOR1, para. 23). In other words, the UNHCR staff's goal was "to directly engage with the COP process and to try and deepen the treatment of human mobility within whatever outcome document is adopted at the end of the negotiations" (IOR1, para. 23). The UNHCR achieved what it wanted as it nowadays has a "hook in the international agreements" (IOR5, para. 67), with paragraph 14(f) of the Cancun Agreement being mentioned repeatedly by IO staff as one big accomplishment, among others, of the UNHCR. Paragraph 14(f) of the Cancun Agreement references migration, planned relocation, and displacement, and IO staff attribute to this paragraph a significant role in influencing states' thinking on matters of displacement and climate change (IOR5, para. 71). The intention was "so that states and donors can say 'okay, they have a taskforce, it could be interesting for us to fund projects' and then the funding comes in" (IOR5, para. 67). That said, the UNHCR wanted, and still wants, to strategically position itself internationally because the global agreements define the strategies for the coming years. IO staff perceive that "having displacement in one of the big agreements will automatically bring states to have some interest in that issue" (IOR5, para. 67).

Therefore, the UNHCR's organizational goal in 2011 was to have states acknowledge that environmental changes will be a new sort of driver of movement. The UNHCR thus "wanted states to adopt a ministerial declaration or a ministerial communiqué" (IOR1, para. 9); in other words, it wanted a global guiding framework on cross-border displacement that the Refugee Convention does not cover, such as climate change and disaster displacement (Goodwin-Gill and McAdam 2017, p. 17). Reflecting an altruistic stance, the organizational goal of the UNHCR has also been described as follows: "Instead of messaging and talking about this abstract group of people who will be moving in the future, why don't we document and help to understand how our people are feeling in these climate change hotspots and what impact we are having in helping them to adapt? And the UNHCR is sitting on a wealth of information in the sense that our clients are on the frontlines of climate change all over the world" (IOR1, para. 18). However, the UNHCR did not accomplish this organizational goal because the states from the global North as well as the global South were reluctant and suspected that the UNHCR's goal was to "refugize the whole area and that somehow the UNHCR was trying to find new clients for the organization" (IOR1, para. 9).

Responding to the criticism, the UNHCR's depiction of the topic happened by two different means. On the one hand, the UNHCR predicted a dark future and hinted at the (national) security implications of EDPs. This depiction is seen, among others, in a 2012 publication in which the UNHCR declared that "environmentally-induced migration and displacement could take on unprecedented dimensions; predictions about the potential scale of such movements range from 25 million to one billion by 2050" (UNHCR 2012, p. 2). This depiction goes back to the organization's internal perception that the topic de facto affects the UNHCR's mandate (IOR5, para. 10). In its yearly publication 'The State of the World's Refugees' of 2012, the organization makes clear that the "UNHCR has a clear interest in such movements of people [EDPs] and an ability to respond to their needs" (UNHCR 2012, p. 2). The organizational goal also becomes apparent when looking at how the UNHCR pushed its own agenda in the development phases of the Copenhagen Agreement (COP 15), the Sendai Framework, and the Paris Agreement (IOR5, para. 67). The organizational goal was to have the parties at the conferences acknowledge the human mobility dimension and "to bring people into their agreements" (IOR1, para. 23).

In 2015, the UNHCR summed up its own agenda on the topic in its publication 'UNHCR, The Environment and Climate Change' as follows: "Rather than calling for a new binding international convention on cross-border displacement, UNHCR supports an approach that focuses on the integration of effective practices by States and (sub-)regional organizations into their own normative frameworks and practices in accordance with their specific situations and challenges" (Türk et al. 2015, p. 7).

#### 5.2.3.2 The Role of the UNHCR Headquarters

(5) The UNHCR headquarters push the organization's EDP approach.

From inside the UNHCR, it is reported that some people at the headquarters had a strong interest in environmental displacement (IOR5, para. 28). As one IO representative explains, "they [people at the UNHCR headquarters] did everything to raise the issue with states and colleagues" (IOR5, para. 28). Despite the influential position of the UNHCR's headquarters on the topic of EDPs, up until 2015, there existed no link between the headquarters in Geneva and the field offices on the issue (IOR5, para. 59). That said, the small group of people working on EDPs at the UNHCR headquarters had to brief the regional offices on the topic, and when the headquarters' staff needed regional information on environmental displacement, they reached out to the UNHCR staff on the ground (IOR5, para. 59). However, this process was still not institutionalized but was an ad hoc process (IOR5, para. 59). In other words, "lessons learned in the field have not been systematically gathered or fed back through to those involved in legal/policy processes" (Goodwin-Gill and McAdam 2017, p. 26). Organization-wise, this means that the knowledge trickled down from Geneva in a top-down manner. Inside the UNHCR, it also depended on whether the regional staff and colleagues on the ground already had experience with earthquakes, flooding, volcanoes, etc. In these cases, the staff quickly tuned in to the headquarters' staff agenda (IOR1, para. 25). Speaking to field office staff, it shows that they largely learned about the topic from the headquarters and were informed that this could be one factor for displacement (IOR2, para. 52). Contrary to this perception, some UNHCR staff felt that the impulse about EDPs had its source from the UNHCR staff working on the ground, because those people worked with refugees and heard about the respective push factors of displacement—among others, disasters and impacts from climate change (IOR5, para. 27).

In the ExCom's 'Draft Report of the 50th Meeting of the Standing Committee of 2011', the UNHCR's role in natural disasters was specifically mentioned (Executive Committee of the High Commissioner's Programme 2011, p. 3). As written in the draft report, the Director of the Regional Bureau for Asia and the Pacific of the UNHCR argued that the region had suffered 70% of the world's recent natural disasters and that, according to his opinion, the UNHCR "could fill an important gap in inter-agency efforts by strengthening partnerships at the country and regional levels in order to make protection support available to governments" (Executive

Committee of the High Commissioner's Programme 2011, p. 5). This shows that the impetus to involve the UNHCR in matters of environmental change not only arose at the headquarters level but was similarly discussed in some regional bureaus. In this context, the UNHCR's dealing with EDPs is reminiscent of the UNHCR's first steps toward opening to IDPs: where a staff member had experienced the situation in Bosnia in the 1990s, this person was sensitized to IDPs (AR1, para. 46). These personnel carried their experience portfolio into the next mission, or even tried to find a follow-up UNHCR operation that also dealt with IDPs (AR1, para. 46).

#### 5.2.3.3 The Role of Individuals at the UNHCR

(6) Individuals in the organization push the UNHCR's EDP approach.

The following staff members filled the leading positions at the UNHCR under the presidency of António Guterres from 2005 until 2015. In 2005 and 2006, Wendy Chamberlin (USA) took the position of Deputy High Commissioner, followed by L. Craig Johnstone (USA) from 2007 until 2009. From 2010 until the end of Guterres' reign, T. Alexander Aleinikoff (USA) filled the position. Hence, the US staffed the influential position of Deputy High Commissioner throughout the whole period of analysis. Concerning the Assistant High Commissioner's position, from 2006 onwards, Guterres installed two senior UNHCR staff members, one for operations and one for protection. The position of Assistant High Commissioner for Protection matters for this analysis because the UNHCR sees IDPs and EDPs from the protection point of view—i.e., the topic falls under this staff member's agenda. From 2006 until her retirement in 2013, Erika Feller (Australia) filled the position of Assistant High Commissioner for Protection. Volker Türk (Germany) took over the post in 2015.

Generally speaking, the organizational structure of the UNHCR illustrates that the number of staff working on this issue is quite manageable. For the timeframe of this analysis, the people who campaigned for the UNHCR's agenda on the issue were the UN High Commissioner for Refugees, António Guterres, the UN Deputy High Commissioner for Refugees, Craig L. Johnstone, and the Assistant High Commissioner for Protection, Volker Türk. Added to this, Marine Franck worked as a consultant at the UNHCR and represented the organization at international conferences and committees, cooperating with states and other IOs on the topic. She was the UNHCR spokesperson and the coordinator of the Advisory Group on Climate Change and Human Mobility (Advisory Group on Climate Change and Human Mobility 2015, p. 9). The hierarchy inside the UNHCR during the period of analysis was that Marine Franck's supervisor, Ellen Hansen, was assistant to the Assistant High Commissioner, Volker Türk, and Volker Türk had direct contact with High Commissioner António Guterres (IOR5, para. 77). This led to a short line of communication from the person working on EDPs, Marine Franck, to the top level of the organization. As a result, when there were urgent matters to discuss, Franck would inform her supervisor, Hansen, who talked to Volker Türk about it. Cooperation inside the UNHCR also took place between Franck and the divisions and people who worked on disaster risk reduction (IOR5, para. 99).

At a conference on Climate Change and Forced Migration organized by the Institute for Public Policy Research in London in April of 2008, the UN Deputy High Commissioner for Refugees made a strong statement concerning the UNHCR's position on EDPs. In his role as Deputy HC, Johnstone argued that "some substantial percentage of the people who will be displaced will be escaping conflict or persecution brought on by civil strife caused in turn by climate change. Those who flee their country because the country no longer exists may well become stateless and therefore become charges of UNHCR. In all of these situations, UNHCR has the mandate and responsibility to assume the responsibility to protect" (Johnstone 2008b, p. 47). On these grounds, Johnstone declared that the UNHCR was often called upon in situations of natural disaster because—even though they do not fall under its mandate—the UNHCR has the capability and the experience to assist these people and even "has a humanitarian obligation in these situations to do what it can" (Johnstone 2008b, p. 47). Also in 2008, Johnstone elaborated further on the humanitarian consequences of climate change. In his remarks, it is seen how Johnstone perceived the role of the UNHCR, as he declared that "I speak to you as a professional devoted to crisis management from the vantage point of the foremost organization in the world dealing with forced displacement, emergency response and recovery from disaster" (Johnstone 2008a, p. 1). This quote is remarkable because the word 'refugee' is not part of it. In Johnstone's opinion, the link between the UNHCR and natural disasters is clear: "We are not mandated to come to the assistance of the victims of natural disasters but time and time again, we have been called upon to play this role, not because it is an implied mandate, but because in the fact of humanitarian suffering those who can help are those who receive the call" (Johnstone 2008a, p. 1).

This quote shows one line of argument that is used by the organization regarding its approach toward EDPs. The argument goes that the UNHCR is, indirectly and passively, pushed toward the topic by external actors. In the same remark, Johnstone mentioned examples of when the UNHCR was called: the tsunami in South East Asia in 2005, in Pakistan in 2006, and in Myanmar in 2008 (Johnstone 2008a, p. 1). Also hovering above this call for help is the humanitarian role that the UNHCR fulfills. Even though this perception does not necessarily reflect how the whole of the UNHCR saw the situation, Johnstone was the voice of the UNHCR as Deputy High Commissioner from 2007 onwards. Following this argumentation, Johnstone concluded that "in the emergencies that will be spawned by climate change, we will be called again and we will have the duty to respond so, by extension, we need to be prepared to carry out this response" (Johnstone 2008a, p. 2). By making this statement, Johnstone was the most outspoken among the UNHCR staff at that time on the topic of UNHCR involvement and he was the only one who went so far as calling it a 'duty' for the UNHCR. Johnstone argued that "more immediately relevant to the UNHCR areas of responsibility are those instances in which climate change strikes in an unpredictable and dramatic way, in the cases of catastrophic weather, or emergency conditions brought on by famine or drought. [. . .] And, of course, UNHCR will continue to manage those circumstances in which climate change precipitates armed conflict and the creation of refugees" (Johnstone 2008a, p. 3). Adding to the assumed duty of the UNHCR in the matter, Deputy High Commissioner Johnstone also called the international community's attention to the issue of preparedness. He stated that "in neither case is the world prepared to cope with this level of human movement. [. . .] I can assure you that the international community is not in a position to respond to the needs of three million people who have been displaced in any given year due to emergency situations" (Johnstone 2008a, p. 3). Johnstone ended his highly alarmist predictions of a negative future with a call for more resources.

Another influential individual on the issue was José Riera, who wrote the first policy statement of the UNHCR on the issue in 2008 (UNHCR 2008). At that time, Riera worked for the UNHCR's policy development and evaluation service, but he volunteered to look into the issue of environmental displacement in order to see how the topic implicated the work of the UNHCR (IOR1, para. 7). Riera stressed that "it has to be acknowledged that the international community has lacked the political will and cooperative spirit required to reduce the pace of climate change" (Riera 2013b, p. 3). At the UNHCR headquarters, José Riera is also called 'Mister Climate Change' by some of his colleagues (IOR2, para. 40). Some of the staff specify Riera as being the person behind the High Commissioner's engagement on the topic because he supported the UNHCR's approach in terms of both content and strategy (IOR2, para. 42). José Riera also closely cooperated with Volker Türk to promote that climate change was indeed a phenomenon and that, among others, it would play out in human mobility terms (IOR1, para. 22). In his publications and statements, Riera remarked that he did not concur with the securitization of EDPs that could be witnessed in the media and in some public statements by politicians. Instead, in 2013, Riera emphasized that "the parade of horrors argument that migration will involve large-scale international movement is simply not supported by existing patterns of movement from natural disasters" (Riera 2013b, p. 2). A clear human security emphasis shows in his declaration that "even if they are not refugees, such people are entitled to be supported and to have their voices heard and taken into account" (Riera 2013b, p. 3).

State representatives also perceived the role of Volker Türk as a leading one in pushing the issue forward. The committed states identified Volker Türk as being "in the driving seat" (SR6, para. 23). When talking to people who worked closely with the above-mentioned individuals, it is hinted that Volker Türk approved of the course of the Nansen process. It is also seen that Türk used human security references in his talking points on EDPs. During a discussion forum on climate change in 2014, Türk enumerated that "the most vulnerable are men, women, boys and girls who also happen to be indigenous and island peoples, nomadic peoples, those living off land and sea with traditions dating back millennia, those already uprooted by conflicts and disasters and needing to pick up stakes yet again or be trapped where they are" (Türk 2014, p. 1). Up until 2015, Türk and other individuals at the UNHCR nevertheless seemed to have realized that their organization was not yet able to push for a larger role in the issue (IOR3, para. 55). One expert sums up this strategy as follows: "to pull the UNHCR in would be a bit like pulling teeth" (IOR3, para. 50). Instead, the organization rather fulfilled its role in the state-led process of the Nansen Initiative. One IO representative argued, "why would you bring this back to the UN when you already have a fantastic group of states?" (IOR3, para. 55).

In addition to the individuals mentioned above, Walter Kälin held a special position among the other actors. Kälin fulfilled the role of negotiator between IOs, between states, and between IOs and states; he functioned as the spokesperson for the Nansen Initiative and as the Chair of the successor to the Nansen Initiative, the Platform on Disaster Displacement. McAdam describes the role of Kälin in the Nansen Initiative as follows: "His personal leadership style of consultation and consensus-building helped to foster understanding and trust, while his clear sense of direction and purpose kept the discussions and outcomes on track. [...] He was mindful of the importance of linguistic precision and the boundaries of normative protection [...] and he understood the proclivities and sensitivities of state actors as well" (McAdam 2017, p. 1523). Walter Kälin made numerous visits to countries which were affected by disasters and the implications of climate change, and he wrote a report on the topic for the UNHCR in 2009 (AR1, para. 3). In 2010 and 2011, Kälin conducted a study for the UNHCR (AR1, para. 6).

In addition to Kälin, Atle Solberg functioned as the chief of the Nansen secretariat and thereby also held a central role among the other individuals. State, IO, and NGO representatives widely expressed their praise for both Kälin and Solberg in their respective roles in getting the issue onto the international agenda (NGOR1, para. 22; SR5, para. 28). One NGO representative assesses that "without Walter Kälin, the Nansen Initiative wouldn't have been the success story it has been. It is really important having him there and then, having someone like Atle Solberg inside the secretariat. They all couldn't have done it without!" (NGOR1, para. 29) Solberg's success in acting behind the scenes has been accredited by the devoted agencies and states, as he was given the central coordinating role on behalf of Germany after July 2016. As Solberg had worked at the UNHCR himself, he is sensitized concerning the different IOs' agendas.

From 2008 onwards, the UNHCR staff began working with the IOM, the IFRC, the UNU in Bonn, the NRC, and with Walter Kälin, back then in his position as Representative of the United Nations' Secretary General on the Human Rights of Internally Displaced Persons (IOR1, para. 7). This development ran parallel to High Commissioner Guterres' first public statements on the matter, which included quite dramatic descriptions of how many people would be displaced by every centimeter rise of sea level (IOR1, para. 7). All the while, the agencies and individuals worked together on behalf of the IASC "to really think through the implications of climate change for the constellation of partners that decided to come together on this" (IOR1, para. 7). Parallel to the IASC collaboration, the individuals at the UNHCR used part of their European Commission grant to fund the Nansen Initiative and the position of Marine Franck; the rest of the fund was put toward a new UNHCR approach to planned relocation (IOR1, para. 29). Under these circumstances, individuals at the UNHCR were able to contribute a lot to international developments on the topic of EDPs, among other things by leading the Global Protection Cluster (IOR3, para. 55). What shows is that the inter-organizational cooperation between the UNHCR and the IOM goes back to the committed staff from both organizations. Based on the initiative of the UNHCR personnel, the Advisory Group on Climate Change and Human Mobility was founded and coordinated by the UNHCR (IOR5, para. 63). As one NGO representative explains, the topic "is a bit of an underdog issue in the disaster and in the displacement world, which means that people who work on it realize that they need each other" (NGOR1, para. 28). The staff at the UNHCR even refer to the experts of the other agencies—e.g., the NRC, the IOM, etc.—as their team, and close colleagues (IOR5, para. 99). This "luck of personalities" (NGOR1, para. 28) pushed people to work together on an issue on which they would normally not have collaborated.

#### 5.2.3.4 The Role of the High Commissioner

(7) The High Commissioner pushes the approach and has a personal motive for doing so.

António Guterres' reign as High Commissioner overlaps with the period of analysis (2005–2015). His follow-up position was United Nations Secretary General from 2016 onwards. Prior to his position as High Commissioner, Guterres was Prime Minister of Portugal and President of the Socialist International. Due to his engagement to respond to the crisis in East Timor and his initiative as president of the European Council in 2000 that led to the Lisbon Agenda, Guterres had proven European and international diplomatic experience (Betts et al. 2012, p. 69). Among other topics, Guterres focused on stateless people, established the Partnership in Action Program Framework to strengthen cooperation with NGOs, and built up the UNHCR Innovation Section (Druke 2013, p. 474). When Guterres took office in 2005, he inherited from his predecessor an organization with financial difficulties. The donor states pressured the organization to reduce the budget at the UNHCR's headquarters in Geneva in order to put more resources into the field operations (Betts et al. 2012, p. 70). Against this background, the new High Commissioner launched a major management reform process by massively reducing the personnel at the headquarters (Betts et al. 2012, p. 70). During his time as High Commissioner at the UNHCR, Guterres stabilized and increased the UNHCR's budget. He also introduced the UNHCR's needs-based budget in 2011 that consisted of US$3.3 billion (Druke 2013, p. 474). Added to this, the organization's budget rose from US $1 billion to US$1.8 billion during Guterres' reign and relations with key host states such as Tanzania and Pakistan improved immensely under his presidency (Betts et al. 2012, p. 78).

Voices from inside the UNHCR confirm that during 2005 and 2015, the topic of EDPs had been picked up at the highest level of the organization—by the High Commissioner (IOR5, para. 10). Also, individuals at the UNHCR assess the High Commissioner's role as highly important when it came to pushing for the topic in international committees (IOR5, para. 31). Inside the UNHCR, the staff who worked on environmental matters during the 2000s still implicate Guterres in the topic of climate change as he was the one who brought the topic to the UN Security Council (IOR2, para. 42). The external experts—state representatives and other IOs' representatives—also attribute the increase in the UNHCR's interest in the topic of EDPs to the role of Guterres (AR1, para. 8). Leading academics in the field of migration and climate change hold that Guterres "seems personally to be driving the institution's engagement with the issue" (McAdam 2011, p. 171) and that "the strong personal conviction of High Commissioner António Guterres in large part drove UNHCR's engagement with the issue" (Goodwin-Gill and McAdam 2017, p. 14).

In his opening statement to the ExCom in 2007, Guterres raised awareness of the topic of forced displacement of people due to environmental degradation and climate change (Guterres 2007, p. 3). Guterres predicted that the long-term effects of climate change would be an expansion of desertification; he also made his (in)famous predication that "for each centimeter the sea level rises, there will be one million more displaced" (Guterres 2007, p. 3). It was not uncommon for him to add to his official talking points, and the above-mentioned claim was made on his own accord (Hall 2015, p. 88). Moreover, even though Guterres stressed in his ExCom statement that "understanding the broader picture should not divert us from our mandate" (Guterres 2007, p. 3), his earlier prediction of one million displaced per centimeter was the part which stuck and which was later criticized. The critique on Guterres' statement comes down to, first, the fact that the information is rather imprecise and, second, that his predication largely called for the attention of states. The High Commissioner's initial engagement with the topic of EDPs goes back to his personal conviction that climate change had developed into a global priority and that the UNHCR should therefore focus on it (Hall 2015, p. 88).

In a similar vein, Guterres declared in his article in 'Foreign Affairs' in 2008 that "new patterns of movement, including forms of forced displacement not envisaged by the Refugee Convention, have emerged" (Guterres 2008, p. 90). Guterres concluded that climate change would worsen human displacement in the future and that "natural disasters that kill and displace millions of people, usually the poorest and the most vulnerable, have become more frequent" (Guterres 2008, p. 99). Drawing on military jargon, Guterres portrays climate change as a force multiplier, because it will lead to displaced people as well as an intensification of other triggers of movement (Guterres 2008, p. 99). The depiction that Guterres draws on in this publication has a securitizing tendency. In this context, he repeats his earlier prediction and refers to former US Vice President Al Gore who estimated that "environmental changes that cause the sea level to rise by one centimeter would displace a million people" (Guterres 2008, p. 99). Guterres concluded his article by declaring that the impact of climate change on human displacement and mobility had so far received too little attention.

Also in 2008, Guterres wrote in a UNHCR publication that "the process of climate change—and the multiple natural disasters it will engender—will in all certainty add to the scale and complexity of human mobility and displacement" (Guterres 2008, p. 1). Guterres explains the role of the UNHCR in reference to climate change being a humanitarian problem and challenge; consequently, the topic of environmental change is of direct interest to humanitarian agencies, including the UNHCR (Guterres 2008, p. 1). Guterres' depiction of climate change contains a lot of security references and is highly dramatic. According to the High Commissioner, the impacts of climate change undermined the security and livelihoods of many people and the occurrence of natural disasters had doubled from 200 to over 400 per year over the last two decades (Guterres 2008, p. 2). In his publication, Guterres differentiated between the large number of people that would be displaced due to sudden-onset events and the large movements of people due to long-term environmental changes. High Commissioner Guterres also depicted the involvement of the UNHCR in cases of internal environmental displacement on a case-by-case basis (Guterres 2008, p. 6). In the same publication, Guterres employed the human security rationale as he stated that "displacement or migration is often the final survival strategy employed by populations whose human security is threatened" (Guterres 2008, p. 9).

Nevertheless, the peak of the High Commissioner pushing the topic is fixed on the year 2011. In June of that year, Guterres gave a speech at the Nansen Conference in Oslo in which he linked climate change with water scarcity, urbanization, population growth, and food and energy insecurity. He argued that "as we look into the future, it seems certain that these trends will increasingly interact with each other, creating the potential for increased competition and conflict over scarce natural resources. As a result, we are also likely to see growing numbers of people being displaced from one community, country and continent to another" (Guterres 2011a, p. 2). Here, the security connotation is visible. This also shows in Guterres' statement that climate change "has the greatest multiplier effect in terms of human security and human mobility" (Guterres 2011a, p. 2). Guterres explicitly mentioned human security in this context and he described in great detail the effects of climate change on human wellbeing and survival. Guterres differentiated between slow- and fast-onset environmental changes and spoke of "large numbers of people" (Guterres 2011a, p. 2) who would be uprooted by natural disasters in a matter of hours, "forcing them to flee for their lives in conditions that resemble refugee movements" (Guterres 2011a, p. 2). By establishing this link between EDPs and refugees, Guterres sensibilized his listeners at the Nansen Conference to a possible involvement of the UNHCR. The High Commissioner offered support to the Nansen Initiative: he proposed a global guiding framework for situations of cross-border displacement due to natural disasters or climate change that states could develop with the help of the UNHCR; also, Guterres proposed that procedures and standards be developed to help the affected populations (Guterres 2011a, p. 4). Guterres notably called on the Nansen Conference participants to develop a set of principles (in this context, he suggested calling them Nansen principles) that could guide discussions with states at the UNHCR's Ministerial Meeting in December of 2011 (Guterres 2011a, p. 5).

In November of 2011, Guterres gave his ‘Maintenance of International Peace and Security’ speech in front of the UN Security Council in which he asserted that “the patterns of movement are changing; meanwhile, the scale of human displacement is also growing” (Guterres 2011b, p. 2). He brought forward the argument that climate change is the challenge of the twenty-first century and “it is a challenge which is adding to the scale and complexity of human displacement; and a challenge that has important implications for the maintenance of international peace and security” (Guterres 2011b, p. 2). In this vein, Guterres linked displacement with climate change and with security concerns and he used the ideational depiction of securitization. Furthermore, the High Commissioner presented four linkages between climate change and displacement: climate change leads to a limited amount of arable land; climate change leads to competition over resources, which might ultimately lead to conflict; climate change leads to more natural disasters; and climate change is a threat to small island states and therefore climate change and citizenship are interlinked (Guterres 2011b, pp. 3–4). According to Guterres, the security impacts of climate change show in his four linkages and he called for a closure of the governance gap that existed in terms of a comprehensive response to climate change, food insecurity, population growth, and water scarcity. Guterres called for “a massive programme of support for adaptation” (Guterres 2011b, p. 5) and “a set of principles, specifically designed to reinforce the protection of and to find solutions for people who have been forced to leave their own country as a result of catastrophic environmental events, and who may not qualify for refugee status under international law” (Guterres 2011b, p. 5).

After the success of establishing paragraph 14(f) at the COP in Cancun in 2010 (AR1, para. 8), “Guterres really tried to institutionalize it as a scope of work for the UNHCR” (IOR3, para. 15). In order to be able to provide input to the IASC Task Force on Climate Change, Guterres created a task force on climate change at the UNHCR with the aim of monitoring developments and finding a consistent approach for his organization (Goodwin-Gill and McAdam 2017, p. 15). Added to this, Guterres chose the EDP topic as the central theme for the UNHCR’s anniversary year in 2011 and “he pushed pretty hard for the topic” (AR1, para. 8). In his ‘Report of the United Nations High Commissioner for Refugees’ of 2013, Guterres repeats his above-mentioned linkage of drivers of displacement and specifically mentions natural disasters and climate change (UNHCR 2013, p. 4). Guterres tried to position the UNHCR before the UN Climate Change Conference in Paris in 2015: as it became clear that there would be a task force for displacement, the High Commissioner specifically asked the Committee to have the UNHCR involved in the issue (IOR5, para. 45). Overall, a number of experts agree that the High Commissioner actually pushed for a bigger role than the one that the UNHCR ultimately received as a standing invitee in the Nansen process (IOR3, para. 55).

Therefore, the empirical analysis presents arguments why the High Commissioner became such an advocate for EDPs: the relevance of the organization and his legacy as High Commissioner. First, some staff members of the UNHCR believe that the High Commissioner’s strong interest can be traced back to the fact that EDPs de facto already affected the UNHCR’s mandate (IOR5, para. 10). That said, Guterres’ personal commitment comes down to the argument that the topic was important from the perspective of protection issues (SR5, para. 18), on which the UNHCR is the leading actor. The interest of Guterres is sometimes ascribed to the wish to make a multi-layered displacement agency out of the UNHCR, including refugees, IDPs, and all those displaced (IOR3, para. 21). Furthermore, because Guterres perceived that climate change was the challenge of this time and age, he saw a role for the UNHCR in this (IOR3, para. 22). Thus, it was Guterres' political judgment that "UNHCR should engage with the issues of the day to stay relevant in an increasingly crowded humanitarian field" (Hall 2015, p. 94). In other words, the High Commissioner had a pragmatic interest in EDPs (Hall 2015, p. 94). Also, Guterres was a strong supporter of the Nansen Initiative, in particular of Walter Kälin, the special envoy of the initiative (AR1, para. 25). Some state representatives interpret this step as a means to expand the organization's work (SR1, para. 44–47). IO representatives and the involved academia speculate that Guterres' commitment might arise from his concern that the UNHCR stayed relevant as an organization (AR1, para. 35). Consequently, he intended to expand his organization into the broad topic of forced displacement.

Second, there are tentative explanations that the High Commissioner pushed the topic of EDPs because at the beginning of his tenure the number of refugees in Europe declined, which means that the UNHCR's case load—refugees—was retrogressive (AR1, para. 25). In other words, Guterres saw an institutional interest in pushing the topic of EDPs onto his organization's radar (AR1, para. 25). Also, having been a politician all his life, UNHCR staff speculate that Guterres "was aware of what issues were increasing in political salience and thus sought to link UNHCR's mandate to it" (Hall 2015, p. 92). Nevertheless, experts do not interpret this as a mere publicity stunt, but say that he was really committed to EDPs and really wanted to include them in the UNHCR's mandate (AR1, para. 25).

Third, some state representatives argue that Guterres intended to maneuver the UNHCR into a leadership position on the topic during his time as High Commissioner (SR1, para. 44). Guterres participated in some of the meetings of the Nansen Initiative, even in meetings where his presence had not been anticipated (SR5, para. 29). The international partners working together in the Nansen Initiative knew that he completely supported the initiative and they stress that in the ExCom statements Guterres always referred to the Nansen Initiative (IOR3, para. 48). If Guterres was not able to attend, the Assistant High Commissioner for Protection, Volker Türk, took part on behalf of the UNHCR (SR5, para. 33). In other words, the organization sent high level staff to discuss EDP matters together with state and IO representatives.

The responses to the High Commissioner's interest in EDP are heterogeneous. Some members of staff are critical about the way that the High Commissioner depicted the topic. These people argue that when speaking about people's lives, the UNHCR always has to be precise and rely on concrete facts (IOR1, para. 8), and even though the future number of EDPs was and is still open to speculation, Guterres' speeches until 2011 gave dramatic predictions of the future number of displaced people (IOR1, para. 7). In other words, even though some approved of the message about linking climate change and forced displacement, the alarmist wording was frowned upon by some staff (IOR1, para. 13).

Table 5.3 depicts the tabular summary of Sect. 5.2.3 and gives an overview of the character of the UNHCR's bureaucracy.

### 5.2.4 *The Influence of Intra-organizational Strategies*

Building upon the focus on the organizational environment, the states, and the character of the bureaucracy, the attention will now shift to the intra-organizational strategies. That said, this section focuses on the following assumptions concerning the UNHCR's approach toward the issue of EDPs (for a summary of this section see Table 5.4):

(8) The EDP approach is a means to enhance the UNHCR's public image.
(9) The UNHCR's EDP approach represents a long-term learning process.
(10) The UNHCR's EDP approach depicts a change in the organization.
(11) The UNHCR approaches the issue because it correlates with its norms and values.

These assumptions will be approached by analyzing the following coding: 'façade', 'organizational learning', 'organizational change', and 'organizational norms'.

#### 5.2.4.1 The UNHCR's Façade

(8) The EDP approach is a means to enhance the UNHCR's public image.

The public image that the UNHCR wants to depict to the international community sometimes is not reflected in the organization's mandate. In other words, the façade of the organization must be taken into consideration to draw conclusions about the organization's strategies.

To start with, the UNHCR depicts itself as a highly transparent and communicative organization. Nevertheless, it seems that up until 2015 there was no link between headquarters and field offices on the topic of EDPs (IOR5, para. 59). Rather, even though the organization tried to depict the image in the early 2010s that it was one of the experts on the topic, just one person at the UNHCR worked specifically on EDPs (though this person interacted with other divisions) (IOR5,

**Table 5.4** The influence of intra-organizational strategies

| Ideational depiction (broadly informed by security theory) | Organizational factors (operationalized organizational theory) |
|---|---|
| **2005–2011** | |
| – Façade: UNHCR poses its EDP involvement as a reaction to external calls for assistance (reactive), even though the organization was also proactive on the topic<br>– Previous organizational change shows in the use of terminology: from refugees, to asylum seekers, to persons of concern, to people on the move in the mid-2000s in Guterres' speeches<br>– **2008**: Deputy HC Johnstone refers to humanitarian obligation of the UNHCR to do what it can to help<br>– **2009**: UNHCR stresses that it is difficult to differentiate between the drivers of displacement; this depiction helps the organization to explain its involvement with EDPs | – **2005**: UNHCR became the leading organization on conflict-induced IDP protection in the cluster approach<br>– **~2008**: Beginning of organizational learning (Guterres' speeches on EDPs); internal task force on climate change; IASC Task Force on Climate Change's sub-group on migration and displacement; first EDP policy paper of the UNHCR; UNHCR's Division for International Protection picks up the topic<br>– Organizational goal to become a recognized international actor on all matters of protection and human mobility<br>– No link between headquarters and field offices on the topic<br>– UNHCR's hierarchical structure and leadership stance influence the organizational culture<br>– Hierarchical predisposition foretells why the UNHCR's top level failed to fully institutionalize the topic in the organization |
| – **Conclusion A**: Organizational culture explains the top-down decision-making of Guterres on EDPs<br>– **Conclusion B**: Organizational goal to raise awareness among states and to stand to benefit from the subsequent funding | |
| **2011–2015** | |
| – **2009**: UNHCR report shows a human security depiction and encourages the international community to adapt a human rights-based approach to migration<br>– **Post-2011**: Development from earlier securitizing tendencies into more empirically-based depictions<br>– UNHCR depicts its EDP involvement as yielding benefits for the fulfilment of its core mandate<br>– **2012**: UNHCR publication forecasts a dark future and hints at (national) security implications of EDPs<br>– **2012**: Book chapter by Guterres conveys that he sees an organizational change which peaked in the UNHCR seeking a more predictable role in its response to EDPs<br>– Conference in **2013**: José Riera uses a human security depiction of EDPs to elaborate on the UNHCR's role; he mentions displacement and migration (shift from the previous migration | – UNHCR public image plays the part of a leading expert on EDPs while there was just one staff member working on the topic<br>– The organization is stretched between its own organizational culture, refugees, host states, donor states, NGOs, and the UN family<br>– Organizational norm of being fact-based and precise: explains the internal reluctance and critical voices toward opening to the vague and broad topic of EDPs<br>– The mandate is one central organizational norm: represents its core values and its organizational culture; UNHCR staff are against renegotiating the Refugee Convention<br>– **2009:** UNHCR invokes its mandate norm to argue that in certain cases, EDPs could be refugees in the traditional sense of the term<br>– Organizational norm of protection (the protection motif): UNHCR argues that EDPs have protection needs similar to refugees<br>– Organizational interest is also triggered by |

(continued)

**Table 5.4** (continued)

<table>
<tr><th>Ideational depiction (broadly informed by security theory)</th><th>Organizational factors (operationalized organizational theory)</th></tr>
<tr><td>avoidance); he refers to a climate change-conflict link to explain the UNHCR's EDP approach<br>– 2013: UNHCR intends to convey the image of a people-centered (human security-centered) humanitarian organization<br>– Since 2013: UNHCR agenda at conferences, meetings, and roundtables was to counter the use of the term 'climate refugee'<br>– Since 2013: Committed staff depict the EDP topic to de facto affecting the UNHCR's mandate<br>– OSCE meeting in 2014: Marine Franck uses a human security focus to describe the protection challenge of EDPs<br>– 2014: Volker Türk argues from a human security perspective at a discussion forum on climate change<br>– 2006–2015: 'UNHCR Global Appeal' publications show no references to environmental change or climate change</td><td>the realization that there is money in the topic<br>– 2010: Breakout session on protection gap of EDPs at the High Commissioner's Dialogue on Protection Challenges<br>– UNHCR applies for European Commission grant which later funds the Nansen Initiative staff position at the UNHCR<br>– The prospect of funding explains the 'beauty contest' between the UNHCR and the IOM<br>– Since 2011: UNHCR acts in the background, but succeeds in remaining a relevant actor on the topic; the committed UNHCR staff perceive that they were the ones who pushed states into the Nansen Initiative<br>– Long line of organizational changes in the UNHCR's work and mandate: from refugees, to asylum seekers, returnees, stateless people and IDPs<br>– Since 2011: Organizational goal to have the parties at COPs acknowledge the human mobility dimension of environmental changes<br>– Only little organizational change in how the UNHCR has tried to understand the implications of natural disasters for its work<br>– UNHCR realizes that the topic must be institutionalized for it to be accepted by staff and partners<br>– Until 2015, the topic is not fully integrated into the UNHCR (expert position externally funded)<br>– No profound organizational change because most staff members do not acknowledge the topic of EDPs as being central for the UNHCR</td></tr>
<tr><td colspan="2">– Conclusion A: Organizational learning: UNHCR learned that it was not wise to connect the Refugee Convention with the issue of EDPs; it learned to differentiate, to use cautious interpretations, and to be careful with future predictions (more credibility)<br>– Conclusion B: No proper organizational change: merely personal commitment of Guterres and a few committed individuals; UNHCR did not evolve an institutional identity on EDPs (due to states' reactions at the ExCom in 2011)<br>– Conclusion C: The international focus on natural disasters led to an opening at the UNHCR that later enabled the discursive opening toward climate change</td></tr>
</table>

para. 99). From the perspective of experts, the UNHCR is "still a bit overstrained in pushing the whole issue forward" (IOR4, para. 58). Therefore, other IOs working in this field perceive that the UNHCR is still not sure whether it should be working on EDPs. Some experts describe the UNHCR's initial steps on the topic as follows: "They [the UNHCR] get into it having no idea, having no subject expertise. And as

an institution, it is totally outside their area of expertise" (NGOR2, para. 28). What is interesting in this context is that the UNHCR framed its involvement as reactive instead of proactive; in a briefing note of April 2009, the UNHCR writes that it "has increasingly been called upon to assist the overall humanitarian effort in the aftermath of a natural disaster" (UNHCR 2009, p. 1). By doing so, the organization conveys the image that external actors pushed the UNHCR into becoming involved.

Concerning the occurrence of the topic inside the UNHCR, the external experts contradict the intra-organizational perception that the UNHCR was the one organization that pushed the topic forward. Contrary to this, some experts argue that the UNHCR merely tagged along on the issue later (IOR4, para. 59) or tried to "proliferate their mandate and grab some more work" (SR1, para. 22). In a Q&A on the UNHCR's Protection Cluster coordination role in natural disasters at the ExCom in February 2011, the UNHCR countered these charges by states. Regarding the impact that its expanded role would have on the UNHCR's core mandate, the organization responded that "past experience has shown a negligible impact on the Office's ability to deliver on core mandate responsibilities. In fact, the reverse has been true, and UNHCR's engagement has yielded benefits for the fulfilment of its core mandate" (Executive Committee of the High Commissioner's Programme 2011, p. 2). Added to this, the organization argued that the UNHCR's new engagement would even help the organization's people of concern because they were part of the response plan. Therefore, the UNHCR presented its expanded role as a win-win situation, combining its mandated responsibility (Refugee Convention) with addressing the emergency situation.

The promise of funding also triggered the UNHCR's interest on the topic (AR1, para. 26; SR1, para. 22). As one state representative describes, "the rivalries among IOs is well-known: it's all about money, individuals, and visibility" (SR5, para. 18). This so-called "beauty contest among IOs" (SR5, para. 19) implies that the IO representatives wanted to link the matter of EDPs to their respective IOs. Some experts who worked closely with the UNHCR and the IOM also argue that it was for the sake of the organization that the staff always said that they did not have enough resources (NGOR1, para. 32). The matter of funding concerns all IOs: one IO representative confesses that "there is money in this [EDPs], we can get additional funding if we work on this issue! We can expand and grow and extend the brand and other stuff!" (NGOR2, para. 26) In its answers to the ExCom's Standing Committee concerning the expanded UNHCR role in natural disasters, the organization made sure to stress that "UNHCR's budget structure ensures that resources for responding to natural disasters would not be diverted from UNHCR programmes for refugees or stateless persons. Funds for refugees (under Pillar I of the budget structure) are firewalled from IDP programmes (Pillar IV), which are project-driven" (Executive Committee of the High Commissioner's Programme 2011, p. 4). The UNHCR's work in cases of natural disaster will be funded through inter-agency flash appeals or targeted appeals and the "UNHCR is therefore able to spend only as much as it can raise for the particular project concerned" (Executive Committee of the High Commissioner's Programme 2011, p. 4).

Some state representatives criticize that the UNHCR deliberately left it open whether a person is displaced by conflict (i.e., under the UNHCR's mandate) or displaced by environmental changes (SR1, para. 37). As Volker Türk of the UNHCR put it, "Interestingly, most observers agree that, with some notable exceptions, it is generally difficult to single out climatic and environmental factors as the sole, unilateral driver of population movements" (Türk 2014, p. 2). According to state representatives, this openness allowed the UNHCR to say it is difficult to differentiate between these two categories and therefore they grabbed the work (SR1, para. 37). By this means, the opening up to EDPs was a strategic step by the organization to enhance its status among the other actors of the international community (SR1, para. 44). A UNHCR policy paper of 2009 concludes that "some movements likely to be prompted by climate change could indeed fall within the traditional refugee law framework, bringing them within the ambit of international or regional refugee instruments, or complementary forms of protection, as well as within UNHCR's mandate" (UNHCR 2009, p. 6). Shortly before the COP in Copenhagen in 2009, the UNHCR established that it wanted to play a part in the handling of EDPs. In the policy paper, the organization gave the example of refugee movements due to armed conflict rooted in environmental factors and the organization argued that this example would call for UNHCR involvement (UNHCR 2009, p. 6). A second example of a possible UNHCR engagement was the situation where a government withheld assistance to the victims of natural disasters to punish or marginalize these people. Here, the UNHCR also saw legitimate grounds to involve itself because "the people concerned could legitimately be refugees in the traditional sense of the term" (UNHCR 2009, p. 7).

After the negative feedback from the ExCom in 2011, the UNHCR's approach to the matter was acting in the background (IOR2, para. 26). In this manner, the UNHCR was able to stay relevant on this newly emerging topic (NGOR2, para. 28). As one IO representative explains, "what better mechanism can they [the UNHCR] have on this issue than that states lead it and they [the UNHCR] are not at the front! Because it gives legitimacy to deal with it, but they [the UNHCR] are not really pushing it because they [the UNHCR] can say there are states that deal with it" (IOR3, para. 48). The UNHCR's relevance in the international community is mentioned by various experts who worked closely with or in the organization in Geneva (AR1, para. 26; NGOR2, para. 28). Whilst countries such as Norway and Switzerland became the frontrunners in the Nansen Initiative, the UNHCR helped the states a lot (SR6, para. 23). In the context of the Nansen Initiative, individuals at the UNHCR convey the image that they were actually the ones who pushed states toward the topic and that they personally played an important role in the success of the initiative (IOR1, para. 11). With the help of the newly established Protection Agenda, the UNHCR is nowadays able to refer to this in its concrete activities (SR6, para. 31).

At a conference in 2013, José Riera of the UNHCR stressed that the organization's concern was what kind of impact climate change would have on human beings (Riera 2013a, p. 4). His ideational depiction is clearly on human security, as Riera elaborated that the intention was to make individuals resilient enough to remain in their homes and to strengthen the adaptation of populations. Also, Riera stressed that the UNHCR's concern was that the human rights of those who are displaced—within or across borders—were safeguarded and guaranteed (Riera 2013a, p. 4). In the same presentation, Riera repeatedly mentioned migration and not only displacement (Riera 2013a, p. 5). The earlier reluctance of the UNHCR to add the migration aspect of global environmental change to the calculation is no longer existent.

#### 5.2.4.2 Organizational Learning Inside the UNHCR

(9) The UNHCR's EDP approach represents a long-term learning process.

The thesis at hand takes up the definition of 'learning' by Haas: "The ultimate purpose is redefined, as means as well as ends are questioned [and the] behavior changes as actors question the original implicit theories underlying programs and examine their original values" (Haas 1990, p. 3). An 'organizational learning' by the UNHCR staff thus implies that, in the context of EDPs, the staff recaps earlier decisions. That said, in order to be able to speak of a learning process by UNHCR staff, it would need active implementation of the topic in the respective divisions. A simple briefing of the media personnel and the regional offices does not fulfill the criteria for a learning process.

**The Three Phases of Organizational Learning at the UNHCR**

The learning process inside the UNHCR dates back to the early 1990s, when the organization started to incorporate the environmental impact of refugee camps on the host country's resources such as water, land, and firewood. In this context, the question arose as to how this impact of the UNHCR's population could be extenuated (IOR1, para. 7; IOR2, para. 13). In the 1990s, the UNHCR had already addressed the issue of resilience against climate change at the operational level to some extent (IOR5, para. 16), but implementation of the topic at the policy level is more recent.

Following this initial phase of organizational learning, the topic of IDPs constituted the second phase. The definition of who is an IDP includes natural disasters as one source or trigger of internal displacement (Economic and Social Council, United Nations 1998, p. 5). The UNHCR had dealt with IDPs before, but in a rather ad hoc and pragmatic manner because the organization did not have official or international backing for this (AR1, para. 19). The analysis of the UNHCR indicates that the topic of IDPs is relevant when looking at the matter of EDPs; in other words, it needed the opening to IDPs beforehand (AR1, para 16). Already in 1994, the ExCom declared that it "encourages the UNHCR to continue its efforts [...] to reinforce and structure coordination through existing interagency mechanisms [...] in order to improve the response by the international community to the plight of the internally displaced" (Executive Committee of the High Commissioner's Programme 1994, p. 3). The

UNHCR's involvement in IDP matters therefore goes back to the inter-agency approach of the UN's humanitarian reform of 2005, as well as the improvement of the humanitarian response model by means of the IASC transformative agenda in 2011 (Türk 2014, p. 4). Volker Türk, in his position as Director of International Protection at the UNHCR back then, sums it up as follows: "At the global level, UNHCR leads the protection cluster in both conflict-induced and natural disaster humanitarian crises, and co-leads shelter and camp coordination and camp management clusters in conflict situations. In situations of natural disasters, leadership of the protection cluster is determined by the relative in-country capacities of UNHCR, the OHCHR, and UNICEF" (Türk 2014, p. 4). In 2005, the UNHCR took on the IDP portfolio (IOR3, para. 19). That said, the UNHCR's opening to IDPs was the precursor to its EDP approach and it needed the IDP opening at the UNHCR for states to see that, in the twenty-first century, it is not just the mandate of 1951 that is of concern for the UNHCR (IOR5, para. 36). This organizational development is a sign of organizational learning as per definition.

Nevertheless, this shows that internal development on the topic of environmental changes was rather slow. The third phase of organizational learning only started in around 2008 when High Commissioner Guterres started to pick up the topic of EDPs in his speeches (IOR1, para. 7). The UNHCR established an internal task force on climate change and led the IASC Task Force on Climate Change's sub-group on migration and displacement (Goodwin-Gill and McAdam 2017, p. 15). In 2008, the UNHCR also published its first policy paper on climate change and displacement, focusing on protection strategies, operations management, and advocacy (Goodwin-Gill and McAdam 2017, p. 15). In the same year, the UNHCR started to extensively cooperate with the IOM, the IFRC, the Climate Centre in the Netherlands, with Walter Kälin (at that time, in his role as the Representative of the Rights of IDPs), and with the UNU in Bonn (IOR1, para. 7). In this third phase, the UNHCR's Division for International Protection also picked up the topic of disaster displacement as part of its scope of work (IOR3, para. 6). After around 2010, more and more sections at the UNHCR began to acknowledge that the topic of environmental change was important and the staff noted that funding was coming in (IOR5, para. 43). The combined UNHCR data of the biggest concentrations of refugees and IDPs and the likely climate change hotspots worldwide showed that "our [the UNHCR's] clients are on the frontlines of climate change all over the world. And instead of messaging and talking about this abstract group of people who will be moving in the future, why don't we document and help to understand how our people are feeling in these climate change hotspots and what impact we are having in helping to adapt and to be ready for the foreseeable disasters?" (IOR1, para. 18).

**'Staff Learning' on the Topic of EDPs**

In the beginning, the staff at the UNHCR were skeptical and reluctant on the EDP topic, owing to the issue lying outside the organization's core mandate (Hall 2015, p. 87). The Refugee Convention had an almost constitution-like standing among UNHCR staff, which led to the topic being merely interpreted as a global trend with no specific influence on the UNHCR's work (Hall 2015, pp. 86–87). Nevertheless, intra-organizational awareness of the topic of EDPs changed over the period of analysis. One IO representative recalls that nowadays "there is more and more understanding that it is not just another issue, but it is all very interlinked with what we are already doing in terms of conflict displacement" (IOR5, para. 15). Hereby, some of the staff at the UNHCR underwent a learning process. This evolution manifested itself in the fact that the organization committed to the new international approaches made through the COP process and the Nansen Initiative. At the operational level, the UNHCR staff increasingly linked their own issues to environmental changes (IOR5, para. 45). Some even describe this process as a 'whole system approach' at the UNHCR, with the aim of 'greening the house' (IOR5, para. 45). Others see staff learning in the fact that the committed individuals sent around so-called climate change readers to colleagues, including the latest information and publications on the topic (IOR1, para. 22). These personal initiatives, briefings, and newsletters set in motion the UNHCR's staff reflection on the implications of environmental changes for the organization. The underlying motive and hook was to raise awareness that "these people may end up being part of our concern—either as IDPs or as people who have crossed a border" (IOR1, para. 22). In this context, the question arises as to whether this staff learning process happened throughout the whole organization or whether this was merely the individual perception of some committed staff members.

On the one hand, when speaking to UNHCR staff that did not primarily focus on issues of environmental migration and displacement, critical voices can be heard. In a recent qualitative study on the topic, the researchers summed up their UNHCR interviews by saying that "while individual staff recognized that people could be displaced for multiple reasons, and that displacement drivers were often tightly interconnected, this does not appear to be widely understood across UNHCR as a whole" (Goodwin-Gill and McAdam 2017, p. 26). Therefore, it is questionable whether the events at the UNHCR from 2005 to 2015 can really be described as a general staff learning process. Rather, the individuals who worked on the topic during that timeframe briefed the other departments and regional offices (IOR5, para. 51). In order to be able to speak of a learning process by staff, it would need active implementation of the topic in the respective divisions. The lack of a learning process that includes more than just a few UNHCR personnel shows in the following example: in 2008 and 2009, the committed individuals still had to justify why they intended to go to climate change conferences on behalf of the UNHCR (IOR1, para. 23). Even though the involvement of the UNHCR post-2009 in international committees and its collaboration with other agencies was on the rise, the topic of EDPs only slowly trickled down until 2015 and ultimately manifested itself in the UNHCR's strong role in the Nansen Initiative.

In spite of this, the official engagement of the UNHCR from 2009 onwards can be traced back to only a small number of committed individuals (see Sect. 5.2.3). An example of the repeated involvement of a small group of people at the UNHCR is the expert consultation in San Remo in 2014 on Planned Relocation, Disasters and Climate Change, which was organized by the UNHCR, the Brookings-LSE Project on Internal Displacement, and Georgetown University. Volker Türk, José Riera, and

Marine Franck from the UNHCR were involved in organizing the consultations and in creating the final report on this consultation (UNHCR 2014, p. 3). Added to this, the UNHCR's staff also adapted their approach to dealing with states after the disappointing experiences during the Ministerial Conference of 2011 (see Sect. 5.2.2). Following the San Remo consultation in 2014, there was a meeting in Bellagio in 2015 at which the participants drafted guidance for states. This shows that the UNHCR's staff focused on information gathering and dissemination to states by means of meetings, conferences, publications—e.g., the special issue of the 'Forced Migration Review'—and side events at COPs (Goodwin-Gill and McAdam 2017, p. 21). The commitment of the organization's staff reached its peak in 2015. The UNHCR staff helped to bring "important language on human mobility in the Sendai Framework for Disaster Risk Reduction 2015–2030, the 2030 Agenda for Sustainable Development, and the 2015 Paris Outcome on climate change" (Goodwin-Gill and McAdam 2017, p. 22). The same personnel had also worked together with states and the IOM in the Nansen Initiative and therefore took part in the Global Consultation in October 2015. Additionally, the staff organized a session on climate change and natural hazards during the High Commissioner's Dialogue on Protection Challenges in December 2015; according to Goodwin-Gill and McAdam, this shows that "there has been an increasing degree of organizational interest and diversity of activities" (Goodwin-Gill and McAdam 2017, p. 22).

On the other hand, a learning process by staff does manifest itself regarding the way the UNHCR publicly linked displacement and environmental changes. The UNHCR publications in the period from 2005 to around 2011 show examples of securitizing the issue—for example, by repeatedly stressing the large number of people that would 'invade' Europe in the future due to extreme weather events. In this context, the High Commissioner's speech in front of the ExCom in 2007 was simultaneously a challenge and a surprise for staff (Hall 2015, p. 88). However, the publications from 2011 to 2015 are much more balanced and show a willingness to be fact-based—e.g., by quoting the research results of partnering organizations such as the NRC, the IDMC, or the IOM. As an example from 2015, the publication 'UNHCR, the Environment & Climate Change' states that "the dynamics of the interaction of mobility with climate change are multifaceted and motivations are complex. Environmental conditions [. . .] are just some among the many reasons why people move. Climate change impacts may play a role in these decisions, but direct causation is difficult to establish" (Türk et al. 2015, p. 5).

**'Organizational Learning' in the Depiction of EDPs**

In the High Commissioner's Dialogue on Protection Challenges in 2010, the UNHCR organized a breakout session to discuss the challenge that existed of the lack of international protection concerning people displaced due to climate change and natural disasters (UNHCR 2010, p. 6). As pointed out in an internal document of 2010, states made clear that they had no intention of revisiting the UNHCR's mandate in the context of environmental migration and displacement (Goodwin-Gill and McAdam 2017, p. 16). Still, the UNHCR looked into the topic of planned relocation and applied for a grant from the European Commission that was ultimately used to fund the Nansen Initiative and the climate change expert position at the UNHCR (IOR1, para. 29).

In the post-2011 period, the UNHCR's ideational depiction of the topic of EDPs had changed. The securitizing tendencies around 2008 and 2009 shifted toward a more empirically-based depiction. From 2013 onwards, the UNHCR's agenda for the various conferences, meetings, and expert roundtables was to clarify the terminology—i.e., that the term 'climate refugee' does not exist and should be avoided. In 2014, Volker Türk of the UNHCR classified that most of the environmental migration and displacement would be internal, which implied that the displaced people could not be called refugees because this would require a cross-border movement (Türk 2014, p. 3). The UNHCR rather preferred the term 'external displacement' in the context of cross-border movements driven by environmental factors (Türk 2014, p. 3). At a meeting of the OSCE in 2014, Marine Franck's talking points also focused on case studies (e.g., Typhoon Haiyan in the Philippines in 2013) to explain and outline the impacts of climate change. She depicted the effects on the livelihoods of people, kept a human security focus and stated that "climate-related human mobility will take different forms and will require a variety of responses, especially the protection of the most vulnerable, at all levels" (Franck 2014, p. 1).

This organizational development shows that the UNHCR learned that it was not feasible or wise to connect the Refugee Convention with the issue of EDPs. Instead, the organization learned to differentiate and ground its statements on empirical facts. Volker Türk argued that "harmful action or inaction by a Government in dealing with climate-related events is related to one or more of the Refugee Convention grounds (such as denial of humanitarian assistance to a minority group) and could, as a result, be considered persecution" (Türk 2014, p. 3). The careful weighing up and cautious interpretation ultimately led to more credibility as opposed to states and other international actors. Still, the organization did not fully pick up the recommendations made by leading academics on the topic which, already in 2011, suggested that the UNHCR could provide practice-based and evidence-based research for UNFCCC parties and could support the process with adequate information (Warner 2011, p. 19).

#### 5.2.4.3 Organizational Change Inside the UNHCR

(10) The UNHCR's EDP approach depicts a change in the organization.

Building upon Sect. 4.2.4, this thesis' understanding of 'organizational change' implies that, first, there have been incidents that paved the way to the UNHCR's interest in EDPs and, second, that one can identify long-ranging and distinct changes inside the organization's bureaucracy due to these initial incidents.

In the past, the question of how to stay relevant for states has triggered a change in the UNHCR's work and mandate. The geographical expansion of its workload in 1967 led to a stronger role of delivering material assistance to refugees in the 1980s (Betts 2010, p. 30). Since the 1990s, organizational change has even exceeded the original boundaries of the refugee regime: the 1990s were characterized by UNHCR's bigger role in humanitarian relief and the 2000s saw an organization that became more and more active in the area of migration (Betts 2010, p. 30).[2] Former senior UNHCR staff and proclaimed refugee expert Jeff Crisp saw an organizational change and extension from refugees to asylum seekers, returnees, stateless people, from IDPs to people displaced by natural disasters (Crisp 2009, p. 73).[3] In the 2000s, Crisp observed a UNHCR transformation toward an organization that looked like an "Office of the High Commissioner for Forced Displacement" (Crisp 2009, p. 74). Betts (2010) argues that this organizational change originated from the growing complexity and the increasingly competitive institutional environment. This allowed states to decide with which IO they would like to work and which collaboration best reflected the states' national interest. Ultimately, the "UNHCR has therefore gradually expanded into the areas in which it faces institutional competition" (Betts 2010, p. 30), and this extension also shows in the UNHCR's language, which developed from mandatory use of the terms 'refugee' and 'asylum seeker' toward 'persons of concern' throughout the mid-1990s until the mid-2000s (Crisp 2009, p. 74). From the mid-2000s onwards, the reference to 'people on the move' was on the rise in Guterres' speeches, UNHCR publications, and public statements (Betts et al. 2012, p. 77).

**The UNHCR and IDPs**

The UNHCR has been through different phases over the last decades and these phases do indeed constitute a change in the organization. The previous extension of

---

[2]High Commissioner Sadako Ogata was the first to establish a relationship between refugees and the environment as early as 1992 and she criticized that this relationship had long been overlooked (Ogata 1992). In 2002, the UNHCR published on refugees and the environment, including both the effects of refugee camps on the environment and environmental change as a trigger to become displaced (UNHCR 2002).

[3]By looking at the earlier changes that the UNHCR went through regarding its 'population', important inferences can be drawn to understand the most recent approach to EDPs. Stateless people were officially assigned as part of the UNHCR's responsibility in 1975 and this section of the organization's work is nowadays a well-funded and important scope of work (Crisp 2009, p. 74). The realization of the International Conference on the Protection Mandate of UNHCR in 1998 illustrates that the UNHCR itself was aware of the questions arising due to its ever-increasing workload. During the conference, UNHCR and NGO staff debated whether the organization's humanitarian activities reflected its mandated protection role (Working Group on International Refugee Policy 1999, p. 397). And the then head of the UNHCR's Evaluation and Policy Analysis Unit, Jeff Crisp, asserted that the organization in fact turned toward the topic of reintegration of displaced populations up until 2000 (Crisp 2001, p. 5). According to Crisp, the UNHCR transformed throughout the 1990s from being refugee-specific and reactive to becoming a broadly-based humanitarian agency with a rather proactive and holistic orientation (Crisp 2001, pp. 6–7). On the topic of its urban refugee policy, the UNHCR also expanded its agenda away from the camp model toward an integrative and community-based model (Ward 2014).

the UNHCR's work was in 2005, when the UNHCR, within the framework of the cluster approach, became the leading organization for IDP protection, camp coordination, and emergency shelter. This decision by the IASC goes back to a call from the UN GA for a more predictable humanitarian system (UNHCR 2007, p. 1). The UNHCR framed its own intentions regarding IDPs as follows: "In line with its expertise and experience, and given its strong commitment to become a predictable and dependable partner in responding to the plight of the internally displaced, UNHCR agreed to assume the lead role for the clusters of protection, emergency shelter and camp coordination and management for conflict-induced IDPs" (UNHCR 2007, p. 2). Hereby, the UNHCR made sure that the new engagement with IDPs was not perceived as arising from inside the organization. Instead, the UNHCR argued that its IDP responsibility went back to states' affirmations through ExCom conclusions and UN GA resolutions (UNHCR 2007, p. 2). It also interpreted its role as provider of the last resort in cases where the UNHCR's partners were unable to act and where protection gaps remained (UNHCR 2007, p. 9). The organization also stressed that its involvement with IDPs must not hinder people from crossing borders and seeking refuge in another country (UNHCR 2007, pp. 11–12). In the beginning, the additional workload due to the opening to IDPs did not go hand in hand with additional sources of funding. Therefore, many people inside and outside the organization were critical of this step (Betts 2010, p. 31), not least because this change in the organization's 'persons of concern' touched upon the sovereignty of states (Dingwerth et al. 2014, p. 196).

Jeff Crisp, during his position as head of the Policy Development and Evaluation Service at UNHCR, published on the UNHCR's evolving perspective in 2008. Crisp analyzed the opening of the UNHCR toward the context of international migration, even though the organization had always been reluctant to acknowledge or discuss international migration before (Crisp 2008, pp. 3–4). According to Crisp, this organizational opening also shows in the introduction of the cluster approach, because "a debate is now taking place with regard to the extent that the organization should also play a role in situations where people are obliged to move as a result of climate change, environmental degradation [. . .]" (Crisp 2008, p. 8). In this context, Crisp also quotes the UNHCR position paper of 2008 in which the organization acknowledged that its mandate had been progressively extended to persons in refugee-like situations (Crisp 2008, p. 8).

**The UNHCR's Rising Interest in EDPs**

As regards the most recent discussion at the UNHCR concerning its people of concern, none of the yearly 'UNHCR Global Appeal' publications from 2005 until 2015 include any reference or contain the words 'environmental change' or 'climate change'.[4] This indicates that either the staff who worked on the EDP topic failed in implementing the issue more broadly in the organization, and thus in the official publications, or the rest of the organization did not acknowledge the topic as being important enough to be included in the 'UNHCR Global Appeal'. In 2006, the

[4]The publications only mention 'natural disasters' in the context of the IOM's field of work in the cluster approach.

organization launched the '10-Point Plan for Action on Refugee Protection and Mixed Migration'. By means of this plan, the UNHCR intended to "assist governments in developing protection-sensitive migration policies and solutions for the different categories of people in mixed movements" (Betts et al. 2012, p. 78). Again, the wording is very general and circumnavigates any reference to the environment.

The yearly UNHCR publication 'UNHCR Global Report' and the respective chapters on IDPs from 2007 until 2016 take a different stance when looking at the search terms 'environmental change', 'climate change', and 'natural disasters'. In the 'UNHCR Global Report' of 2008, the UNHCR illustrates that "major gaps remain in guidance on protection in natural disasters. UNHCR [...] embarked in 2008 on a revision of the 'Pilot Field Manual' accompanying the 'IASC Operational Guidelines on Human Rights and Natural Disasters'; this work will be finalized in 2009. Standard operating procedures for the establishment of a protection cluster in disasters are being drafted and will be rolled out to Humanitarian Country Teams in 2009" (UNHCR 2009, p. 50). In the 'UNHCR Global Report' of 2009, the organization declared that in natural disaster situations, the "UNHCR demonstrated its readiness to lead field protection clusters by deploying a protection team to the Philippines" (UNHCR 2010, p. 48). The UNHCR's message was clear: it mentioned the other humanitarian agencies like UNICEF, UNDP, and OCHA, but also highlighted its frontrunner position on the topic and its willingness to do more. In the 'UNHCR Global Report' of 2010, the UNHCR declared that in comparison to conflict-IDPs and violence-IDPs, "millions more had been uprooted by natural disasters" (UNHCR 2011, p. 49).

That said, there is some organizational change if one looks at, first, how the organization has tried better to understand the implications of natural disasters for its work and, second, how the UNHCR has demonstrated that its clients sit on the frontlines of climate change (IOR1, para. 25). In terms of the size and scope of the organization, some experts argue that the UNHCR changed as much as possible within its own means (IOR1, para. 25). From this perspective, the above-mentioned organizational learning relating to IDPs is an organizational change, because the path directly led from IDPs to EDPs and there existed a chronological rigor in these two topics for the UNHCR (AR, para. 16; IOR3, para. 25). What really did shift in the context of organizational change is that the UNHCR established the protection cluster in those countries in which the organization already had missions. This cluster approach also included the topic of natural disasters (AR1, para. 10). Prior to the occurrence of the topic of EDPs, the UNHCR had involved itself with disaster-IDPs, even though this happened on a case-by-case basis (UNHCR 2009, p. 7). Some examples of the organization's emergency intervention in cases of disaster-IDPs are the 2004 Indian Ocean tsunami, the 2005 South Asian earthquake, the 2006 floods in Somalia, the 2007 floods in Pakistan, the 2008 floods in Yemen, the earthquake in China in 2008, and the cyclone-related flooding in Myanmar in 2008 (UNHCR 2009, p. 2).

In 2012, Guterres declared that there had been an evolution of the UNHCR's responsibilities. He summed this evolution up as follows: "From the 'good offices' concept that allowed assistance to individuals and groups not covered by the 1951 convention definition, to our lead responsibility for reducing and preventing statelessness, and a more predictable role responding to people displaced internally by conflict and natural disasters" (Betts et al. 2012, p. xxi). This shows that the High Commissioner saw an organizational change throughout the years and perceived the step toward EDPs as being one of these organizational changes. Still, Guterres made sure to stress that, in his opinion, the additional responsibilities "have in no way undermined our core mandate" (Betts et al. 2012, p. xxi). In other words, he was well aware of the existing criticism concerning the organization's ever-evolving appetite for including more persons of concern in the UNHCR workload.

Some researchers have outlined the UNHCR's expansionist history in terms of geographical scope and member states and interpreted these developments as mirroring the expansionist culture of the UNHCR (Crisp 2009, p. 75). However, when considering all factors, the approach to EDPs is in itself not a proper organizational change. The interest in EDPs goes back to, among others, "the personal commitment of the High Commissioner Guterres, but it never made its way institution-wise or organization-wise; it was very much based on a few people" (IOR3, para. 14). The reason for this lack of institutional identity in relation to EDPs goes back to how the international community (e.g., states) reacted to the UNHCR's initial steps on the topic in 2011. To put it differently, "how consolidated is that [environmental displacement] within the UNHCR? Nothing in the ExCom, no UN GA resolution, very weak policy papers!" (IOR3, para. 37) Based on these facts, a mere briefing of the other divisions and of the media people in the organization does not constitute an organizational change as such, because the change did not reflect institutionally or on the policy level. For example, the UNHCR did not create a new position for the climate change expert in the timeframe of this analysis; instead, the EU, and later Germany as the lead of the Nansen Initiative, funded Marine Franck's position (IOR5, para. 85). Even though the UNHCR had the operational background, most states did not see a role for the UNHCR in the topic and therefore the funding of states did not come in.

#### 5.2.4.4 The UNHCR's Organizational Norms

> (11) The UNHCR approaches the issue because it correlates with its norms and values.

The UNHCR's self-conception as an organization is that, in publications and speeches, the underlying facts must always be precise because it is usually people's lives that the UNHCR refers to (IOR1, para. 8). This explains why the UNHCR is skeptical about the terminology of climate or environmental refugee: "these terms have no basis in international refugee law" (UNHCR 2009, p. 8). One expert claims that "if a refugee comes over the border, it is quite straightforward: you screen them, status determination, you build a camp. [...] In IDP situations, it is much messier and the UNHCR struggles much more" (IOR3, para. 59). Added to this intended precision, the self-conception of the UNHCR is that the Refugee Convention, and thus the mandate, is the basis for its work (IOR2, para. 60). According to IO representatives, "there is a very strong traditional culture and they [UNHCR staff] are very committed to that" (IOR3, para. 27) and this culture is that "the UNHCR has a mandate for refugees, there is a convention and they are excellent on that; the legal understanding and their policy program, it's brilliant!" (IOR3, para. 27). This organizational norm explains why there were a lot of critical voices concerning EDPs and why there was not much progress organization-wise on the topic (IOR3, para. 27). However, some IO representatives perceive the Refugee Convention to be a "living instrument and its shelf life depends on the ability to be flexible to interpret it and to adapt it to changing circumstances" (IOR1, para. 9). However, there exists overall agreement that the UNHCR's mandate is the basis of its organizational culture and it represents the organization's central values (IOR2, para. 60). Another pivotal organizational norm that guides the UNHCR is the age-gender-diversity approach when it comes to the organization's planning process (IOR12, para. 60).

Moreover, the organization's hierarchical structure and the leadership's stance also influence the UNHCR's organizational culture. A unique feature of the organization (next to the OHCHR) in the UN family is that the UNHCR is the bureaucracy as well as an individual person, the High Commissioner (McKittrick 2008, p. 14). In fact, most High Commissioners guided the organization by means of their strategies, ideas, and expertise and used their leadership skills to "alter the information and value contexts in which states made policy" (Loescher and Milner 2011, p. 194). Throughout its existence, the UNHCR has tried to further its normative agenda and the High Commissioners have oftentimes succeeded in negotiating with states so that if they defined their national interests in respect of refugee needs, the states would simultaneously help to ensure inter-state and domestic stability and profit from international cooperation (Loescher and Milner 2011, p. 194). The fact that UNHCR staff are distributed all over the world in over 100 regional and country offices with joint headquarters in Geneva also influences the self-perception of the organization and of the individual staff members. Therefore, it can be assumed—depending on the individual composition of the departments—that some UNHCR staff feel more connected to the regional office as opposed to the headquarters, or that some individuals working in the field feel closer to their operational partners (i.e., fellow IO workers). This also results in autonomous action and "in many cases, officials in 'the field' feel that they have a certain level of autonomy in interpreting the mandate in the context of the local place and situation" (McKittrick 2008, p. 15). As the organization expanded its scope of work, it also grew in terms of staff and budget, which ultimately resulted in frequent rotation and larger divisions. Added to this, the organizational culture of the UNHCR evolved over time as a result of the organization's opening toward new people of concern. In the context of the organizational culture, some researchers question the existing mechanisms of accountability because "external examination of UNHCR and formal mechanisms of oversight, which would hold it accountable for its actions, are insufficient and scarce"

(McKittrick 2008, p. 15). This is problematic on account of the UNHCR being stretched between its own organizational culture, refugees, host states, donor states, NGOs, and the UN family (McKittrick 2008, p. 16).

The topic of natural disasters led to an opening at the UNHCR that later enabled the discursive opening toward environmental changes more generally, including climate change (IOR3, para. 19 and 25). This shows in the UNHCR's normative depiction of the topic as early as 2008, when the leading individuals in the organization already ascribed the issue to the UNHCR on humanitarian grounds. As UN Deputy High Commissioner for Refugees, L. Craig Johnstone, promulgated at a climate change conference, the "UNHCR has the humanitarian obligation in these situations to do what it can" (Johnstone 2008b, p. 47). Johnstone even went to far as to proclaim that "it is hard to envisage a true international emergency of forced displacement precipitated directly or indirectly by climate change in which UNHCR would not play a key role" (Johnstone 2008b, p. 47).

As regards the organization's approach toward EDPs, the UNHCR invoked its mandate as the organizational norm. First, in a policy report of August 2009, the UNHCR closely linked EDPs to its organization's scope of work. The policy report implied that "there may also be situations where the victims of natural disasters flee from their homeland because their government has consciously withheld or obstructed assistance in order to punish or marginalize them on one of the five grounds set out in the refugee definition" (UNHCR 2009, p. 7). Under these circumstances, the UNHCR proclaimed that EDPs "could legitimately be refugees in the traditional sense of the term" (UNHCR 2009, p. 7). This policy report of 2009 shows that the UNHCR brought up the norm of protection. Here, the UNHCR reasoned that "the crux of the issue will be whether persons have a need for international protection; and, if so, on what grounds this need may be turned into an entitlement" (UNHCR 2009, p. 5). The report came out shortly before the COP in Copenhagen in December 2009. Therefore, this depiction of the importance of the UNHCR's role in the matter can be seen from the background of the advocacy work in preparation for the COP. Second, the policy report predicted a future role for the UNHCR in EDP matters and largely traced this role back to the UNHCR's mandate. It is argued that if the inhabitants of the Maldives and Tuvalu need to leave their country as a result of rising sea levels and flooding, the UNHCR will become involved to assist those who seek safety abroad with regard to statelessness (UNHCR 2009, p. 7).

The UNHCR report of 2009 argued that "since new forms and patterns of movement are emerging, the concepts traditionally used to categorize different types of movement are becoming increasingly blurred" (UNHCR 2009, p. 3). Therefore, it proposed that "new legal frameworks may need to be negotiated" (UNHCR 2009, p. 3). However, the UNHCR made clear that this did not imply an opening of the Refugee Convention of 1951: "The UNHCR considers that any initiative to modify this definition would risk a renegotiation of the 1951 Refugee Convention, which would not be justified by actual needs. Moreover, in the current political environment, it could result in a lowering of protection standards for refugees and even undermine the international refugee protection regime altogether"

(UNHCR 2009, p. 9). The UNHCR's staff as well as the states in the ExCom were against any form of opening of the existing legal framework. The inclusion of IDPs under the workload of the UNHCR had already been a highly controversial step internally. In this context, it shows that the depiction of the UNHCR was clearly from a human security perspective. In its report in 2009, the UNHCR called for and encouraged the international community in an approach that was based on respect for human rights and on international cooperation. Furthermore, the organization explicitly mentioned human security in its report when it outlined that migration can be a survival strategy and a form of adaptation (UNHCR 2009, p. 10).

Another organizational norm that guided the organization's decision-making was the protection motif (SR5, para. 18). The protection cluster was established in 2006 and the UNHCR originally interpreted its scope of work as only incorporating conflict-IDPs (IOR3, para. 14). As Volker Türk put it: "We in UNHCR are deeply concerned about the massive protection challenges raised by climate- and environment-related displacement and planned relocation" (Türk 2014, p. 2). Türk also stressed that protection was central in responses to disaster scenarios (Türk 2014, p. 4), and he referred to case study research of the UNHCR of 2013, which "found that many of the protection concerns in disaster contexts were the same as in conflict situations" (Türk 2014, p. 5). In other words, "if you think about displacement really from the logical perspective, it is never just so simple as it is just a conflict or just a disaster" (NGOR2, para. 40). It took some time for the UNHCR to understand that, being the global lead of the protection cluster, the organization could not ignore situations of disaster, even if its mandate did not fully cover them (IOR3, para. 14). In the protection cluster, the role of global lead was separated from the country level: the cluster lead at the country level might be either conflict or disaster displacement and it was either UNICEF or the UNHCR that led at this level (IOR3, para. 14). Besides these two levels (global and country), the UNHCR also had to decide what to do as an operational agency—i.e., concerning its program activities. As one expert sums up, "you can perfectly well decide not to deal with disaster situations from an organizational point of view, but you still may have the role of national cluster lead and also your role as global lead is indisputable" (IOR3, para. 14).

These various levels of operational involvement also triggered the institutional notion of the UNHCR to put the topic of environmental displacement on the organization's radar (IOR3, para. 14). As a result, the UNHCR had to understand the topic of protection from three different perspectives: the situation of armed conflict, the situation of generalized violence and human rights violations, and the situation of disasters (IOR3, para. 14). From a global point of view, the UNHCR needed to build the capacity to do this, even if this ultimately did not lead to a real change in organizational attitude and merely to secondment and consultants (IOR3, para. 14). As Marine Franck of the UNHCR summarized at an international conference on human mobility, the UNHCR involved itself because "the affected populations may have protection needs and vulnerabilities similar to those whose flight is provoked by violence or human rights abuses" (Franck 2015, p. 6). Again, the protection motif became central in this case.

Another example of the organization's own depiction of the topic is the presentation that José Riera made at an international conference in 2013. Like Marine

Franck, Riera also tackled the question of the UNHCR's mandate in the context of EDPs. He acknowledged that the organization's mandate clearly did not include EDPs and presented the following line of argument: "Environmental degradation can fuel social tension and, in some cases, conflict which, in turn, can give rise to flows of refugees and IDPs. Even where the cause of displacement—whether internal or cross-border—is primarily environmental, the affected populations may have protection needs and vulnerabilities similar to those whose flight is provoked by violence or human rights abuses" (Riera 2013a, p. 10). As a result, Riera stated that "the organization [the UNHCR] has a clear interest in the movement of people prompted by these factors [natural disasters and climate change]" (Riera 2013a, p. 10). Next to Riera, Marine Franck also picked up the question of why the UNHCR was concerned with the matter of EDPs in her remarks at a preparatory meeting of the OSCE in 2014. Franck argued that, first, the UNHCR was concerned with climate-related human mobility because the persons of concern to UNHCR "are huddling in most of the major climate change hotspots around the globe" (Franck 2014, p. 1) and, second, the organization was already involved in large-scale disaster response, such as after the Philippines Typhoon Haiyan.

In other words, during the period of analysis, the UNHCR repeatedly connected the topic of EDPs to its organizational norm of protection. The ideational depiction of the issue was related to human security—i.e., it was all about the human side of environmental migration and displacement. As Franck put it, the "UNHCR is particularly concerned about the enjoyment of human rights by people relocated or displaced by the effects of disasters and climate change, either within their own borders or across borders. The affected populations may have protection needs [. . .] similar to those whose flight is provoked by violence or human rights abuses" (Franck 2014, p. 1). Similarly, at a discussion forum on climate change in 2014, Volker Türk argued from a human security perspective. He criticized that from a protection perspective the affected communities were not adequately consulted throughout the relief effort (Türk 2014, p. 5). According to Türk, the problem with disaster response in 2014 was that it often only focused on the immediate life-saving concerns and forgot to incorporate other protection risks, such as gender-based and sexual violence, and separated children (Türk 2014, p. 5).

Table 5.4 depicts the tabular summary of Sect. 5.2.4 and gives an overview of the influence of intra-organizational strategies.

### 5.2.5 *Recapitulating the Analysis*

António Guterres was faced with a number of challenges when he took office in mid-2005. In the first decade of the twenty-first century, the UNHCR was deeply concerned about the restrictive asylum policies of Northern states; this "shrinking asylum space" (Betts et al. 2012, p. 69) led to ever smaller numbers of asylum seekers in industrialized countries. Simultaneously, the organization witnessed an increase in the number of IDPs and saw the occurrence of extraordinary natural disasters, which also led to massive internal displacement. These external challenges on the organization ultimately led to the organization becoming involved in natural disasters (Betts et al. 2012, p. 69).

Nevertheless, inner organizational struggles concerning the UNHCR's new operational outlook arose on the topic of IDPs. From the end of the last century to the beginning of the 2000s, the new protection realities challenged the UNHCR. The scope of its work repeatedly included people displaced due to natural disasters or internally displaced. Instead of the Refugee Convention, these people rather fell under the Guiding Principles on Internal Displacement, which specifically mentions natural disaster as a reason for displacement. Experts credit the Guiding Principles with a broadness that helps to apply them to cases of disasters caused by climate change (Kälin 2010, p. 87). Before being fully implemented in the UNHCR, the policy papers and strategies on IDPs went back and forth between the organizational units and the ExCom multiple times (IOR3, para. 21). Inside the organization, the staff intensely debated the step toward IDPs and there existed an in-house resistance with reference to the strict mandate of the UNHCR (AR1, para. 13). Broadly speaking, an internal debate is still going on among UNHCR staff about IDPs and this in-house division goes so far that some staff talk about refugees even though they deal with IDPs (AR1, para. 46). One example of this is the Africa Bureau at the UNHCR, which is still very refugee-oriented. The underlying reason is that African countries are comparatively tolerant about receiving refugees and largely follow a refugee camp logic so that they can delegate the operations to IOs. By contrast, IDPs are highly political and often go along with conflicts of interest (AR1, para. 46).

The struggles inside the UNHCR can also be traced back to the challenges that arose due to post-9/11 international developments in Iraq and Afghanistan, which resulted in a peak of UNHCR involvement. Around 2009 and 2011, another peak level in terms of refugee flows and IDPs occurred and some IO representatives argue that the UNHCR staff "are hardly able to deal with their mandated responsibility" (IOR3, para. 30). Looking at the inner organizational struggles, it can be seen that the various organizational units depend highly on the financial contributions of the ExCom. The organization has different pillars. The first pillar is about refugees and stateless persons—i.e., the persons of concern as stated in the UNHCR's mandate. The organization is bound to fill this first pillar—the mandatory responsibility—with the ExCom money and the purchase committee controls that the organization acts accordingly (Executive Committee of the High Commissioner's Programme 2011, p. 7). The second pillar is about all other matters of concern to the UNHCR that are not directly linked to the mandate, and this pillar is only filled if more money comes in (Executive Committee of the High Commissioner's Programme 2011, p. 7). In other words, in case of strenuous UNHCR operations that tie up a lot of resources, such as the Syria crisis, the second pillar runs dry of money. Applying this to the topic of EDPs, one IO representative claims that "states should at least recognize that if you want the UNHCR to do that [EDPs], you have to give them the money they need; if you want them to do more, you need more earmarked money for that" (IOR3, para. 40).

Another inner organizational struggle shows in how the UNHCR's institutional structure is built. There exists an environmental unit at the UNHCR with a focus on, among others, the impact of refugee camps on the environment. Nevertheless, this operational unit does not incorporate the effects of environmental changes on human mobility. What shows is that the environmental unit of the UNHCR has not taken the topic of EDPs as part of its operation. One reason for this neglect is that the UNHCR management argued to keep focusing on the environmental management (IOR2, para. 35). That said, there is a discrepancy between the people who work at the environmental unit, who acknowledge that one could approach environmental change and mitigation by means of environmental activities, and the UNHCR's management that did not urge a stronger overlap between these different units (IOR2, 35). Still, the environment and the climate change unit has started to cooperate internally in the last couple of years and has also involved itself externally—for example, in the Green Climate Fund. Inside the UNHCR, a steering committee emerged that is cross-sectoral and that includes the people working on disaster risk reduction, health, and livelihood (IOR2, para. 38; IOR5, para. 99).

Concerning the topic of EDPs, the internal organizational struggles also arose because the UNHCR's management—i.e., the High Commissioner—introduced the topic of EDPs in a top-down manner to his staff. In 2008, the UN Secretary General assigned the task to all agencies in the UN family to set up focal points on climate change in preparation for the COP in Copenhagen in 2009; still, "the office holder [of the newly established climate change position] received no formal briefing note, and the role of focal point was subsequently handed around the organization like a 'hot potato' to three different people in different divisions over three years" (Hall 2015, p. 89). By the same token, the UNHCR staff did not actively back Guterres' task force on climate change and, because the topic was added to their usual workload, the chosen staff in the various divisions did not extensively fulfill their task to develop the organization's position on climate change (Hall 2015, p. 89). In relation to the top-down decision-making of Guterres concerning EDPs, his senior managers did not agree with their role because they "were reluctant to delegate staff to it [the topic of climate change], as they did not see the issue as a core concern" (Hall 2015, p. 89). Herein lies one of the reasons why some staff opposed the topic being included on the organization's agenda (AR1, para. 51). The reluctance of the UNHCR staff was reflected in the organization's behavior from 2007 until 2010, shown by the weak internal support of the climate change focal points, no organizational interest in attracting financing from the climate funds, nor any substantive displacement projects focusing on environmental changes (Hall 2015, p. 91). Another reason is that the Assistant High Commissioner, Erika Feller, always had a focus on refugees and was not very open to focusing on IDPs or on EDPs.

To conclude, this chapter has presented the analysis of this thesis by means of the conceptual design of Chap. 4. Therefore, the focus of this chapter first lay on the role of the organizational environment; second, on the role of states; third, on the character of the bureaucracy; and, fourth, on the organizational strategies. The sections thereby analyzed these factors in terms of their respective explanatory power concerning the UNHCR's EDP approach since 2005. It is seen that a combination of factors played a role in the UNHCR's decision-making from 2005 until 2015 and Chap. 6 will subsume and interpret these findings in a detailed manner.

## References

Advisory Group on Climate Change and Human Mobility (2014) Joint submission to the United Nations Convention on Climate Change (UNFCCC) on National Adaptation Plans (NAPs): for consideration at SBI 40. Advisory Group on Climate Change and Human Mobility. https://unfccc.int/sites/default/files/149.pdf. Accessed 13 June 2018

Advisory Group on Climate Change and Human Mobility (2015) Human mobility in the context of climate change UNFCCC - Paris COP - 21: recommendations from the advisory group on climate change and human mobility. www.unhcr.org/protection/environment/565b21bd9/human-mobility-context-climate-change-unfccc-paris-cop-21-recommendations.html. Accessed 13 June 2018

Barnett MN, Finnemore M (1999) The politics, power and pathologies of international organizations. Int Organ 53:699–732

Betts A (2009) Protection by persuasion: international cooperation in the refugee regime. Cornell University Press, Ithaca, NY

Betts A (2010) The refugee regime complex. Refug Surv Q 29:12–37

Betts A (2011) Introduction: global migration governance. In: Betts A (ed) Global migration governance. Oxford University Press, New York, pp 1–33

Betts A, Loescher G, Milner J (eds) (2012) UNHCR: The politics and practice of refugee protection, 2nd edn. Routledge, London

Calmy-Rey M (2012) Statement delivered at the UNHCR Ministerial Meeting of December 2011. In: Pledges 2011: Ministerial Intergovernmental Event on Refugees and Stateless Persons, Geneva, 7–8 Dec 2011

Crisp (2001) Mind the gap! UNHCR, humanitarian assistance and the development process. Working paper No. 43, New issues in refugee research series. UNHCR Evaluation and Policy Analysis Unit, Geneva. http://www.unhcr.org/research/working/3b309dd07/mind-gap-unhcr-humanitarian-assistance-development-process-jeffcrisp.html. Accessed 13 June 2018

Crisp J (2008) Beyond the nexus: UNHCR's evolving perspective on refugee protection and international migration. New Issues in Refugee Research

Crisp J (2009) Refugees, persons of concern, and people on the move: the broadening boundaries of UNHCR. Refuge 26:73–76

Deschamp B, Azorbo M, Lohse S (2010) Earth, wind and fire: a review of UNHCR's role in recent natural disasters. UNHCR Policy Development and Evaluation Service, Geneva

Dijkzeul D, Beigbeder Y (2003) Rethinking international organizations: pathologies and promise. Berghahn Books, New York

Dingwerth K, Blauberger M, Schneider C (2011) Postnationale Demokratie: Eine Einführung am Beispiel von EU, WTO und UNO. Grundwissen Politik, vol 47. Springer VS, Wiesbaden

Dingwerth K, Lehmann I, Reichel E, Weise T, Witt A (2014) Many pipers, many tunes? Die Legitimationskommunikation internationaler Organisationen in komplexen Umwelten. In: da Conceição-Heldt E, Koch M, Liese A (eds) Internationale Organisationen: Autonomie, Politisierung, interorganisationale Beziehungen und Wandel. Nomos, Baden-Baden, pp 186–212

Druke L (2013) Innovations in refugee protection: a compendium of UNHCR's 60 years: including case studies on IT communities, Vietnamese boatpeople, Chilean exile and Namibian repatriation. Peter Lang Edition, Frankfurt am Main

Economic and Social Council, United Nations (1958) Establishment of the executive committee of the programme of the UNHCR: ECOSOC resolutions, E/RES/672 (XXV). Economic and Social Council, United Nations, New York

Economic and Social Council, United Nations (1998) Further promotion and encouragement of human rights and fundamental freedoms, including the question of the programme and methods of work of the commission: human rights, mass exoduses and displaced persons. Report of the representative of the Secretary-General, Mr. Francis M. Deng. Economic and Social Council, United Nations, New York. https://academic.oup.com/rsq/article/18/1/165/1574028. Accessed 13 June 2018

Entwisle H (2013) The world turned upside down: a review of protection risks and UNHCR's role in natural disasters. New issues in refugee research series. United Nations High Commissioner for Refugees, Policy Development and Evaluation Service's, Geneva. http://www.refworld.org/docid/5142d4652.html. Accessed 13 June 2018

Executive Committee of the High Commissioner's Programme (1994) Internally displaced persons: No. 75 (XLV) – 1994, Executive Committee 45th session. United Nations High Commissioner for Refugees, Geneva. http://www.refworld.org/docid/3ae68c434.html. Accessed 13 June 2018

Executive Committee of the High Commissioner's Programme (2011) Standing committee 51st meeting: UNHCR's role in support of an enhanced humanitarian response for the protection of persons affected by natural disasters. United Nations High Commissioner for Refugees, Geneva. http://www.unhcr.org/4df08ba39.pdf. Accessed 13 June 2018

Ferris E (2014) Planned relocations, disasters and climate change: consolidating good practices and preparing for the future. Background document, Sanremo Consultation, 12–14 March 2014. UNHCR, BROOKINGS and Georgetown University, Geneva. http://www.unhcr.org/53c4d6f99.pdf. Accessed 13 June 2018

Franck M (2014) Cross-dimensional impacts of natural and man-made disasters, population movements in the context of natural disasters. In: Second preparatory meeting of the 22nd OSCE

Franck M (2015) When leaving your home is the last option for survival: climate change and human mobility, Presentation. COST Action 1101 Climate Change and Migration; Knowledge, law and policy, and theory, Durham, UK

General Assembly (1993) A/RES/48/116: 85th plenary meeting. United Nations, New York. http://www.un.org/documents/ga/res/48/a48r116.htm. Accessed 13 June 2018

General Assembly (1994) A/49/12/Add.1: ExCom on IDPs, No. 75 (XLV), 45th session. United Nations, New York. http://www.unhcr.org/excom/exconc/3ae68c434/internally-displaced-persons.html. Accessed 13 June 2018

Glover T, Russell S (2001) Overseeing the refugee convention: coordination with UNHCR and states. ICVA working paper No. 7. International Council of Voluntary Agencies, Geneva. https://papers.ssrn.com/sol3/papers.cfm?abstract_id=294181. Accessed 13 June 2018

Guterres A (2007) Opening statement by Mr. António Guterres, UNHCR: Fifty-eighth Session of the Executive Committee of the High Commissioner's Programme, Geneva

Guterres A (2008) Millions uprooted: saving refugees and the displaced. Foreign Aff 87:90–99

Guterres A (2011a) Nansen conference on climate change and displacement; Statement by António Guterres. UNHCR, Oslo

Guterres A (2011b) Maintenance of international peace and security: new challenges to international peace and conflict prevention, New York

Guterres A (2012) Opening statement. In: Pledges 2011: Ministerial Intergovernmental Event on Refugees and Stateless Persons, Geneva, 7–8 Dec 2011

Haas EB (1990) When knowledge is power: three models of change in international organizations. University of California Press, Cambridge, MA

Hall N (2015) Money or mandate?: Why international organizations engage with the climate change regime. Glob Environ Politics 15:79–97

Hüfner K (2015) Mehr Verantwortung übernehmen: Zum deutschen Finanz-Engagement in den Vereinten Nationen 1991–2013. Frank & Timme, Berlin

Informal Group on Migration/Displacement of the IASC (2008) Climate change, migration and displacement: who will be affected? Working paper. Inter-Agency Standing Committee, Geneva. http://www.unhcr.org/protection/environment/4a1e4fb42/climate-changemigration-displacement-affected-working-paper-submitted.html. Accessed 13 June 2018

Inter-Agency Standing Committee (2011) IASC operational guidelines on the protection of persons in situations of natural disasters. The Brookings - Bern Project on Internal Displacement, Washington

Johnstone CL (2008a) Planning for the inevitable, the humanitarian consequences of climate change: remarks by Ambassador L. Craig Johnstone, UN Deputy High Commissioner for Refugees

Johnstone CL (2008b) The future is now. Forced Migr Rev 31:47

Kälin W (2010) Conceptualizing climate-induced displacement. In: McAdam J (ed) Climate change and displacement. Hart, Oxford, pp 81–104

Kälin W, Schrepfer N (2012) Protecting people crossing borders in the context of climate change: normative gaps and possible approaches. Legal and protection policy research series. UNHCR Division of International Protection, Geneva. http://www.unhcr.org/4f33f1729.pdf. Accessed 13 June 2018

Kimani MJ (1996) Bridging relief and development concepts in refugee and migration situations: household energy programmes. In: IOM (ed) Environmentally-induced population displacements and environmental impacts resulting from mass migrations: International symposium, Geneva, 21–24 Apr 1996, pp 115–120

Loescher G, Milner J (2011) UNHCR and the global governance of refugees. In: Betts A (ed) Global migration governance. Oxford University Press, New York, pp 189–209

Loescher G, Betts A, Milner J (2012) The UNHCR: The politics and practice of refugee protection, 2nd edn. Routledge, New York

McAdam J (2011) Environmental migration. In: Betts A (ed) Global migration governance. Oxford University Press, New York, pp 153–188

McAdam J (2017) From the Nansen initiative to the platform on disaster displacement: shaping international approaches to climate change, disasters and displacement. UNSW Law J 39:1518–1546

McConnachie K (2014) Governing refugees: justice, order and legal pluralism. Routledge, New York

McGee J, Steffek J (2016) The Copenhagen turn in global climate governance and the contentious history of differentiation in international law. J Environ Law 28:37–63

McKittrick A (2008) UNHCR as an autonomous organisation: complex operations and the case of Kosovo. Working paper series No. 50. University of Oxford

Mearns R, Norton A (eds) (2010) Social dimensions of climate change: equity and vulnerability in a warming world. The World Bank, Washington, DC

Ministerial Communiqué (2011) Intergovernmental event at the ministerial level of member states of the United Nations on the occasion of the 60th anniversary of the 1951 convention relating to the status of refugees and the 50th anniversary of the 1961 convention on the reduction of statelessness, 7–8 December 2011. United Nations High Commissioner for Refugees, Geneva. http://www.refworld.org/docid/4ee99bf42.html. Accessed 13 June 2018

Nansen Conference (2011) Nansen conference on climate change and displacement in the 21st century. Oslo, 6–7 June 2011, Summary. Norwegian Refugee Council/Internal Displacement Monitoring Centre, Oslo. http://www.refworld.org/docid/521485ef4.html. Accessed 13 June 2018

Ogata S (1992) Statement by Mrs. Sadako Ogata at the UN conference on environment and development. Rio de Janeiro, 10 June 1992. United Nations High Commissioner for Refugees, Geneva. www.unhcr.org/admin/hcspeeches/3ae68fcd4c/statement-mrs-sadako-ogata-united-nations-high-commissioner-refugees-united.html. Accessed 13 June 2018

Organisation of African Unity (1969) OAU convention governing the specific aspects of refugee problems in Africa. Organisation of African Unity. https://www.refworld.org/docid/3ae6b36018.html. Accessed 2 July 2019

Pease K-KS (2016) International organizations, 5th edn. Routledge, New York

Riera J (2013a) Challenges relating to climate change induced displacement. In: International conference "Millions of people without protection: climate change induced displacement in developing countries", Berlin

Riera J (2013b) Seminar to address the adverse impacts of climate change on the full enjoyment of human rights: international cooperation and respect for human rights. International Cooperation and Respect for Human Rights, Palais de Nations, Salle XII, UNHCR headquarter Geneva

Schorr M (2004) Der Wandel der humanitären Aktion internationaler Organisationen: Die institutionellen sowie materiell-rechtlichen Konsequenzen dargestellt am Beispiel des IKRK, UNHCR und UNHCHR. Verlag Dr. Kovac, Hamburg

The Nansen Initiative (2017) Towards a protection agenda for people displaced across borders in the context of disasters and the effects of climate change. The Nansen Initiative, Geneva. https://www.nanseninitiative.org/secretariat/. Accessed 13 June 2018

Türk V (1999) The role of UNHCR in the development of international refugee law. In: Nicholson F, Twomey P (eds) Refugee rights and realities: evolving international concepts and regimes. Cambridge University Press, Cambridge, pp 153–175

Türk V (2014) Discussion forum on climate change: remarks by Volker Türk, Director of International Protection, UNHCR Geneva, Berlin

Türk V, Corliss S, Riera J, Lippmann B, Hansen E, Egziabher AG, Franck M, Dekrout A, Kuroiwa Y (2015) UNHCR, the environment and climate change. Updated version. UNHCR. http://www.unhcr.org/540854f49.pdf. Accessed 13 June 2018

UN General Assembly (2009a) Follow-up to the outcome of the Millennium Summit: 85th plenary meeting, 3 June 2009, A/63/PV.85. UN General Assembly, New York. http://www.securitycouncilreport.org/atf/cf/%7B65BFCF9B-6D27-4E9C-8CD3-CF6E4FF96FF9%7D/CC%20A%2063%20PV.85.pdf. Accessed 13 June 2018

UN General Assembly (2009b) Climate change and its possible security implications: Report of the Secretary-General, A/64/350. UN General Assembly, New York. http://www.unhcr.org/protection/environment/543e73f69/climate-change-its-possiblesecurity-implications-report-secretary-general.html. Accessed 13 June 2018

UN General Assembly (2009c) Resolution adopted by the General Assembly: climate change and its possible security implications. UN General Assembly, New York. https://www.un.org/en/ga/search/view_doc.asp?symbol=A/RES/63/281. Accessed 13 June 2018

UN Office of the High Commissioner for Human Rights (OHCHR) (1993) Fact sheet No. 20: human rights and refugees. OHCHR, Geneva. https://www.refworld.org/docid/4794773f0.html. Accessed 2 July 2019

UNFCCC (2011) Report of the conference of the parties on its sixteenth session, held in Cancun from 29 November to 10 December 2010. Decisions adopted by the Conference of the Parties. United Nations. https://unfccc.int/resource/docs/2010/cop16/eng/07a01.pdf. Accessed 13 June 2018

UNHCR (1951) Convention and protocol relating to the status of refugees. UNHCR, Geneva. https://www.unhcr.org/protection/basic/3b66c2aa10/convention-protocol-relatin-status-refugees.html. Accessed 2 July 2019

UNHCR (2002) A critical time for the environment. Refugees, pp 4–14

UNHCR (2007) The protection of internally displaced persons and the role of the UNHCR. Informal consultative meeting. United Nations High Commissioner for Refugees, Geneva. http://www.unhcr.org/protection/idps/50f951df9/protection-internally-displaced-persons-role-unhcr-excom-informal-consultative.html. Accessed 13 June 2018

UNHCR (2008) Climate change, natural disasters and human displacement: a UNHCR perspective. United Nations High Commissioner for Refugees, Geneva. http://www.refworld.org/docid/492bb6b92.html. Accessed 13 June 2018

UNHCR (2009) Internally displaced persons. In: UNHCR (ed) UNHCR Global Report 2008, pp 46–51

UNHCR (2010) Working with internally displaced. In: UNHCR (ed) UNHCR Global Report 2009, Geneva, pp 47–49
UNHCR (2011) Working with the internally displaced. In: UNHCR (ed) UNHCR Global Report 2010, Geneva, pp 48–53
UNHCR (2012) Displacement, climate change, and natural disasters (Chap. 7). In: UNHCR (ed) The state of the world's refugees 2012: in search of solidarity, Geneva
UNHCR (2013) Report of the United Nations High Commissioner for refugees. Part II, Strategic review pursuant to General Assembly resolution 58/153. United Nations, New York. http://www.unhcr.org/excom/unhcrannual/5244440f9/report-united-nations-high-commissioner-refugees-part-ii-strategic-review.html. Accessed 13 June 2018
UNHCR (2014) Planned relocation, disasters and climate change: consolidating good practices and preparing for the future. Expert consultations. Final Report, Sanremo, Italy. Georgetown University, Georgetown. http://www.unhcr.org/54082cc69.pdf. Accessed 13 June 2018
UNHCR (2015) Contributions to UNHCR for budget year 2015. United Nations High Commissioner for Refugees, Geneva. https://www.unhcr.org/partners/donors/558a639f9/contributions-unhcr-budget-year-2015-31-december-2015.html. Accessed 2 July 2019
UNHCR (2017) Governance and oversight. http://www.unhcr.org/governance.html. Accessed 13 Mar 2019
UNHCR, IOM, Refugee Policy Group (1996) Environmentally-induced population displacements and environmental impacts resulting from mass migrations. International symposium, Geneva, 21–24 April 1996. International Organization of Migration, Geneva. http://publications.iom.int/system/files/pdf/environmentally_induced.pdf. Accessed 13 June 2018
UNHCR, IOM, NRC, UNU, Representative of the Secretary General on the Human Rights of Internally Displaced Persons (2009a) Comments and proposed revisions to the negotiating text prepared by the Chair of the UNFCCC Ad Hoc Working Group on long-term cooperative action. United Nations High Commissioner for Refugees, Geneva. https://unfccc.int/resource/docs/2009/smsn/igo/055.pdf. Accessed 13 June 2018
UNHCR, IOM, NRC (2009b) Climate change and statelessness: an overview. Submission to the 6th session of the Ad Hoc Working Group on Long-Term Cooperative Action under the UNFCCC. United Nations High Commissioner for Refugees, Geneva. http://www.unhcr.org/protection/environment/4a1e50082/climate-change-statelessness-overview.html. Accessed 13 June 2018
UNHCR, NRC, Representative of the Secretary General on the Human Rights of Internally Displaced Persons, UNU (2009c) Forced displacement in the context of climate change: challenges for states under international law. United Nations High Commissioner for Refugees, Geneva. http://www.unhcr.org/protection/environment/4a1e4d8c2/forced-displacement-context-climate-change-challenges-states-underinternational.html. Accessed 13 June 2018
United Nations Security Council (2007) 5663rd meeting, Tuesday, 17 April 2007, New York, S/PV.5663. United Nations, New York. http://repository.un.org/handle/11176/9267. Accessed 13 June 2018
Ward P (2014) Refugee cities: reflections on the development and impact of UNHCR urban refugee policy in the Middle East. Refug Surv Q 33:77–93
Warner K (2011) Climate change induced displacement: adaptation policy in the context of the UNFCCC climate negotiations: Research Series. UNHCR Division of International Protection, Geneva. http://www.unhcr.org/4df9cc309.pdf. Accessed 13 June 2018
Warner K, Kälin W, Martin S, Nassef Y, Lee S, Melde S, Chapuisat HE, Franck M, Afifi T (2014) Integrating human mobility issues within national adaptation plans. United Nations University - Nansen initiative joint policy brief No. 2. UN University, Bonn. https://environmentalmigration.iom.int/integrating-human-mobility-issues-within-national-adaptation-plans. Accessed 13 June 2018
Working Group on International Refugee Policy (1999) International conference on the protection mandate of UNHCR – Final Report: Peace Palace, The Hague, The Netherlands, 18 September 1998. Int J Refugee Law 11:397–416

# Chapter 6
# Conclusion: Putting the UNHCR's EDP-Approach into Context

*What I am asking of you here today is not a new convention, it is not an extended mandate. What I am asking is for all of us to assume our shared duty.*
ExCom Ministerial Meeting (Guterres 2012b, p. 145)

This chapter first presents the findings of the analysis. The following sections interpret the findings and discuss the implications of the thesis' findings for security theory, for organizational research, for the research on human mobility and for IR theory. After a critical reflection on the research design and a presentation of the policy implications, the last section elaborates on the way ahead for human mobility in general and EDPs in particular.

## 6.1 Presentation of the Findings

The findings will be presented in the same theory-led manner as the analysis chapter. The following paragraphs subsume the four sections on, first, the organizational environment; second, the member states; third, the character of the bureaucracy; and, fourth, the intra-organizational strategies (for a summary of this section see Table 6.1). Thereby, this section constitutes the first step toward answering the research question:

Why does the UNHCR approach the issue of EDPs the way that it does?



S. Hantscher, *The UNHCR and Disaster Displacement in the 21st Century*, Contributions to Political Science, https://doi.org/10.1007/978-3-030-19689-9_6

**Table 6.1** Tabular summary of the empirical analysis

| Ideational depiction (broadly informed by security theory) | Organizational factors (operationalized organizational theory) |
|---|---|
| 2005–2011 | |
| – The **international environment** played a large role and pushed the UNHCR into the topic of EDPs<br>– The UNHCR learned from the organizational environment that it would be helpful to jump onto the bandwagon of **climate change** (pre-COP 2009) to push its own agenda forward (extend its protection agenda)<br>– The ideational depiction of EDPs shows that natural disasters are not as powerful as an **advocacy instrument** as the buzzwords 'climate change'<br>– **The national interests of states** tie the UNHCR's hands on the EDP topic—no role of states in pushing the organization's preliminary steps on EDPs<br>*Quality of the UNHCR's approach until 2011:*<br>– **No organizational interest** at the UNHCR before 2009; compared with other IOs, the UNHCR allocated few members of its personnel to the EDP topic<br>– **Staff were reluctant** to become involved in the EDP topic<br>– **High Commissioner Guterres** promoted the topic<br>– It was the **organizational goal** to raise awareness among states—and to benefit from the funding | |
| 2011–2015 | |
| – **International environment:** The UNHCR collaborated with other IOs and succeeded in pushing the topic into international forums—it needed the IOs (in collaboration with some committed states)<br>– The UNHCR has to balance the wishes of the global North and the global South<br>– Attitudinal change among states: The **Nansen Initiative states** play a large role in the UNHCR's EDP approach (intention of the Nansen states: to pull the UNHCR closer to the Initiative and to bring the IOM and the UNHCR closer together)<br>– Committed **individuals** at the UNHCR were central in establishing the organization's prominent role in international forums; personal dedication of just a few staff members<br>– **Organizational learning**: the UNHCR learned not to connect the Refugee Convention with the EDP topic; the UNHCR realized that it needs to institutionalize the topic first, before staff and partners accept it as a new topic<br>– By opening up to **natural disasters**, the UNHCR's debates subsequently shifted to **climate change**<br>– The UNHCR's 'public image'- goals: to show a **human-security**-centered humanitarian IO<br>– From 2011 onwards, the depiction concurs with the hidden **UNHCR agenda** to guarantee future funding and to work on its international status and relevance<br>– The UNHCR's depiction of its EDP interest: as a displacement agency, it has a legitimate stance on EDP matters and knows how to deal with the displaced; it follows the moral grounds of protection; to understand all other—non-Refugee Convention—reasons for displacement<br>*Quality of the UNHCR's approach until 2015:*<br>– **UNHCR individuals** succeeded in positioning the topic in international forums, but this engagement was not reflected/mirrored in the organization itself<br>– The topic of EDPs has not been institutionalized; a lack of funding (due to earmarked funding and the pillar-structure at the UNHCR) resulted in a limited amount of resources dedicated to the issue of EDPs<br>– Little to no internal awareness of the UNHCR's work on the topic<br>– Guterres failed to integrate EDPs fully into the UNHCR's work<br>– No proper **organizational change** inside the UNHCR with regard to the topic of EDPs; no new institutional identity | |

Table 6.1 summarizes the findings of the analysis in a concise manner. The table explains the UNHCR's approach to EDPs and delivers answers to the research question of this thesis.

### 6.1.1 The 'Organizational Environment': The Source of the EDP Approach?

Building on the analysis in Sect. 6.1.1, the primary and secondary data indicate that the external environment of the UNHCR has had an influence on the organization's EDP approach. There had been no significant organizational interest at the UNHCR on the matter of EDPs before the COP in 2009, but "the international community's focus on climate change and natural disasters has changed that" (Deschamp et al. 2010, p. 3). The debates in the organizational environment of the UNHCR as well as in the international forums changed enormously between 2005 and 2015—some examples of this development are the international negotiations on environmental changes and the various international forums on human mobility. Concerning the inter-organizational cooperation in the UNHCR's environment, it shows that the issue of EDPs alludes closely to the topic of protection. Being *the* international protection agency, the UNHCR's limited mandate still poses a challenge, because it restricts its organizational cooperation on EDPs.

The issue of EDPs is strongly related to the topic of international migration, but migration was not fully settled or agreed upon in the UN until 2016. Instead of an international migration regime within the UN, the project-dependent IOM only recently—in September 2016—became part of the UN family. Experts in Geneva trace the lack of a UN-migration-regime back to states' caution about an institution that might curtail their sovereignty or even oblige them in their migration admission policies (AR1, para. 56). The data indicate that IOs played a fundamental role, in cooperation with states, in pushing the topic of EDPs in the international forums (SR5, para. 13–14). A statement by High Commissioner Guterres sums up the relationship between the IOM and the UNHCR: "They are twins, but the IOM lost its birth certificate" (SR3, para. 10). Furthermore, the general tendency among the experts is that it needed both the UNHCR and the IOM (NGOR1, para. 20–21). Even though some experts perceive the competition between the IOM and the UNHCR as a "beauty contest" (SR5, para. 19–20), that is, each IO tried to persuade the international community that it is the logical partner on the topic, they do not assess this rivalry as being destructive (SR5, para. 19–20). Rather, the two IOs found common ground on the topic (IOR3, para. 49–50), and both organizations tried to stress their respective advantage regarding the topic of EDPs (SR5, para. 19–20). Therefore, the developments in the organizational environment influenced the UNHCR's interest in EDPs, but no external actor explicitly delegated the organization to the topic.

### *6.1.2 The 'Member States': The Source of the EDP Approach?*

Section 6.1.2 describes the way in which states reacted to the UNHCR's approach to EDPs and gives hints about an attitudinal change among states. Goodwin-Gill and McAdam have the impression that these changes are apparent "in terms of their [states] awareness of the impact of disasters and climate change on displacement, their willingness to engage with it, and their acceptance of UNHCR's role in both operational and standard-setting areas" (Goodwin-Gill and McAdam 2017, p. 23). That said, most of the states did *not* push for or support a UNHCR approach toward EDPs. In this context, the UNHCR finds itself facing a dilemma: "The UNHCR is highly politicized and is limited by states' concerns regarding sovereignty. Because it relies on states for its very existence, UNHCR must defer to the interests of its donors and host governments" (Glover and Russell 2001, p. 5). Nevertheless, a small number of states—those that worked closely together in the Nansen Initiative—collaborated with the UNHCR and the IOM on the topic. This network of experts was made up of state representatives as well as migration and refugee specialists. Furthermore, the Nansen Initiative's work helped to put the topic of environmental migration and displacement on the agenda of more and more states. Therefore, instead of directly initiating the topic of EDPs on the agenda of the UNHCR, states rather tried to pull the organization closer to the Nansen Initiative and to push the IOM and the UNHCR closer together (SR5, para. 26). This is also evident in the follow-up of the Nansen Initiative, the Platform on Disaster Displacement. According to states, the change in the title symbolizes a change in focus from a mere research perspective (Nansen Initiative) toward implementation (Platform on Disaster Displacement) (SR5, para. 10). Hereby, the rise in the global awareness of the topic of EDPs is evident in the Paris Agreement, which was signed by 175 states in April 2016.

Therefore, states did not have a stake in the preliminary step of the UNHCR's approach to EDPs; instead, the Nansen Initiative states strengthened the UNHCR's interest in EDPs on its way to becoming a knowledgeable and welcome invitee to the respective international committees on the topic. After 2011, a small number of states took over the issue and formed a coalition of committed states whilst taking the input and expertise of IOs and NGOs into consideration.

### *6.1.3 The 'Character of the Bureaucracy': The Source of the EDP Approach?*

The analysis in Sect. 6.1.3 concerning the character of the bureaucracy reaches to the following conclusions: the hierarchical structure of the UNHCR did *not* hinder the organization's approach; instead, it depended on the individuals themselves, for example whether they were dedicated and wanted to strengthen the UNHCR's bargaining position on the topic. Hereby, the implementation of new ideas and projects largely depended on questions of funding, because all UNHCR projects need to bear reference to earmarked funding; thus, UNHCR staff members have to follow the money to implement a new idea (IOR5, para. 83). In addition, it shows that the individuals at the UNHCR acknowledged that it is necessary for the organization to focus on EDPs, as they "may end up being part of our concern—either as IDPs or as people who have crossed a border" (IOR1, para. 22).

Nonetheless, the group of individuals who worked on the topic from 2005 until 2015 is manageable, which means that—even though the topic was perceived as important—the group was of a small size organization-wise. In addition, states and other agencies noticed that the UNHCR only devoted a limited amount of resources to the issue of EDPs (NGOR1, para. 27; SR3, para. 41). What is also apparent is that, of the individuals who involved themselves in the issue internationally, a high percentage were staff members of the UNHCR or individuals who had a previous connection to the UNHCR. Though these individuals were able to position the topic in the international committees and among state representatives, this engagement nevertheless was *not* mirrored in the organization itself. Until 2015, the UNHCR had not fundamentally institutionalized the topic of EDPs, and the organization did not grant any additional staffing positions. This means that, around 2013 and 2014, "there were two people, José Riera and Marine Franck, those two, and they had their networks" (NGOR1, para. 27).

This thesis argues that, until 2015, there was little awareness of the topic internally, that is, within the UNHCR. Thereby, the thesis concurs with previous qualitative research on the UNHCR. As Goodwin-Gill and McAdam put it, "those working in operations tend to see it as self-evident that UNHCR should (and does) provide protection and assistance in the aftermath of a disaster [...]. Some working in the legal/standard-setting area are more cautious about UNHCR's work in this area [...]. However, in part, that view seems to be linked with concerns about lack of capacity and resources, rather than with the mandate in a formal sense" (Goodwin-Gill and McAdam 2017, p. 29). That said, the organization is still in the development phase of fully institutionalizing the topic of EDPs in the UNHCR. In comparison with the successful opening toward IDPs, High Commissioner Guterres failed in his agenda to integrate EDPs fully into the UNHCR's work (AR1, para. 36). One reason is the intra-organizational reluctance to embrace the topic and the lack of will of the organization's staff. Another reason is that states clearly signaled that they would not follow Guterres' EDP plans in 2011. Lastly, world politics, specifically the wars in Iraq and Syria, tied up many international resources (AR1, para. 36).

### 6.1.4 *The 'Intra-Organizational Strategies': The Source of the EDP Approach?*

Regarding the analysis of the intra-organizational strategies in Sect. 6.1.4, the thesis comes to the following conclusion: when looking at the period between 2005 and

2015, it shows that the UNHCR went a long way toward realizing that the topic of EDPs has to be institutionalized inside the organization before staff and partners accept the UNHCR's role in EDP protection. An example is the externally funded expert position of Marine Franck in 2013. Many secondments and consultants but no ordinary staff worked on the topic within the UNHCR (IOR3, para. 14), and the UNHCR did not evolve an institutional identity on the issue of EDPs. Therefore, no proper organizational change is visible during the time period from 2005 until 2015 (also cf. Hall 2016). Even though some funding was available until 2015 and more people inside the UNHCR were interested in the topic, the issue was not a top priority when considering the overall agenda of the organization (see Sect. 5.2.4). Hence, the UNHCR's engagement with displacement in the context of environmental changes lacks an institutional memory and "remains to be mainstreamed" (Goodwin-Gill and McAdam 2017, p. 29). Until 2015, the UNHCR had not fully integrated the topic into the organization, and it will still take a few more years for the topic of EDPs to be institutionalized (IOR5, para. 11).

However, from 2011 onwards, the UNHCR wanted to convey a public image of a people-centered (i.e., a human security-centered) humanitarian organization that is emancipated and that also focuses on aspects that lie outside of its realm, such as migration—a topic that falls under the auspices of the IOM. In this manner, the depiction from 2011 until 2015 concurs with the hidden UNHCR agendas mentioned in Sect. 5.2: to guarantee its future funding and to work on its international status and relevance. These reasons played a role in the UNHCR's opening toward EDPs.

Whether one gives credence to this organizational façade or not, experts from academia and from the IO environment in Geneva all accredit great competencies to the UNHCR in matters of protection (IOR3, para. 34). Additionally, some IO representatives believe that, even though "the UNHCR should not get into major long-term population movements such as migration, the sudden-onset disasters are particularly part of the UNHCR's protection agenda" (IOR3, para. 34). To put it differently, the organization has a legitimate stance on EDP matters, "because the UNHCR is a displacement agency, they really know how to deal with the displaced" (IOR3, para. 59). In the initial phases of the UNHCR's approach until the 2010s, "there is no organizational 'culture' of protection in this area" (Goodwin-Gill and McAdam 2017, p. 26). However, throughout Guterres' reign, the "UNHCR sought moral legitimacy" (Hall 2015, p. 94), and the organization's arguments were mostly based on the moral grounds of protection. Hereby, the UNHCR's goals to raise awareness of states and push them into action might date back to the organization's success in responding to natural disaster-related displacement situations throughout the period of analysis (Goodwin-Gill and McAdam 2017, p. 23).

The UNHCR intended to develop the law surrounding refugees and IDPs and to explore planned relocation as a form of adaptation (IOR1, para. 29). That is why the UNHCR conducted empirical research, reporting and capacity building of its population of concern (IOR2, para. 21). From this perspective, the answer to the question of why the UNHCR approaches the topic of EDPs is that the organization's goal was really to understand all the other—non-1951 Refugee Convention—

reasons for displacement, and this interest and eagerness led to research and to the practical involvement of the UNHCR (IOR2, para. 50).

## 6.2 Interpretation of the Findings

By making the arguments presented above, this thesis contributes to and challenges the existing research. It generates implications for four areas of research: security research, organizational research, refugee research and IR research.

### *6.2.1 Implications for Security Research*

The thesis uses the security perspectives as the starting point to outline the empirical puzzle. First, this thesis contributes to recent security research that has identified security references at the UNHCR during the last two decades (Hammerstad 2014, p. 152). The thesis fills in the details on how the UNHCR used security references between 2005 and 2015. Furthermore, it indicates that, from the moment when Guterres took office in 2005 until the year 2011, the concept of securitization of the Copenhagen School and Paris School best captures the ideational depictions used by the UNHCR. As outlined in Sect. 5.2, this is apparent in the way in which the organization's website, the official publications and the UNHCR's speeches presented EDPs. The UNHCR fell in line with the way in which many actors—NGOs, IOs and the media—talked about the nexus of environmental changes (including climate change) and human mobility. Until 2009, many international actors referred to future scenarios of large numbers of displaced people who would overwhelm European countries and to developments that would possibly lead to conflict with local populations and clashes over declining resources in the global South. This thesis argues that the UNHCR picked up this interpretation in 2005 and let itself be directed by the buzzwords 'climate change'.

Additionally, the thesis shows that the UNHCR purposefully turned its attention away from the buzzwords 'climate change' in 2011 and toward the overlying and broader topic of EDPs. This evolution could only be carved out by means of the different lenses of security studies. The analysis indicates that the UNHCR intended to shift toward environmental migration and displacement to avoid the debate about whether a certain natural disaster occurred due to climate change. Instead, the UNHCR emphasized the displaced people, the EDPs, and, with the help of the concept of Human Security, the thesis is able to grasp the changing outlook of the UNHCR. Here, the analysis adds flesh to earlier studies that saw a "more cautious, protection-centered and less ambitious [UNHCR] discourse by 2012" (Hammerstad 2014, p. 152). Some scholars argue that the Human Security depiction of environmental changes helped to politicize the topic (Oels 2012, p. 196). Applying this argument to EDPs, the depiction of the UNHCR after the COP in Copenhagen in

2009 helped to change many of the states' approaches on the topic. This evolution in the international community on the topic is apparent in the development of the Nansen Initiative, with just a few committed states, into the Nansen Initiative Protection Agenda, which 109 governmental delegations endorsed in October 2015.

Overall, this thesis also adds to security research to the extent that it discusses the question of why global actors use traditional security connotations or references to human security. Here, it becomes important that the UNHCR did not use these ideational depictions when talking to its people of concern, for example refugees or IDPs, but instead only used these references toward its 'audience'—that is, its donor states and the international community. To incorporate security research into an organizational analysis therefore helps to gain insights into the organizational strategies and goals. Nevertheless, the Copenhagen School, the Paris School and the Human Security concept themselves do not shed light on the question of why the UNHCR acts in this way. Consequently, this thesis does not contribute to theory building in security research but shows how the 'lenses' of security research capture an empirical puzzle and thereby represent the starting point of an in-depth organizational analysis of a 'puzzling' empirical observation. Guterres did not refer to security to locate the UNHCR among the international security actors (e.g. NATO); rather, "instead of a global security actor, Guterres' aimed to make UNHCR into the UN's 'protection agency'" (Hammerstad 2014, p. 161). This thesis uses the theoretical 'lenses' of security research as guidance to approach the empirical puzzle and to capture the changes in the UNHCR's use of security references. However, security theories hereby constitute the preliminary steps, and the thesis does not apply security theories in a discourse-analytical manner.

Nevertheless, based on the above-mentioned arguments, security research has proved to be relevant to the research purpose of this thesis. In addition, recent research also calls for a careful consideration of matters of security in research on environmental changes and climate-conflict research (Adams et al. 2018; Baldwin et al. 2014; Boas and Rothe 2016; Kita and Raleigh 2018; Oels and von Lucke 2015; Rothe 2016; Trombetta 2014). In this context, researchers need to be aware of the conclusions that might be drawn from their research—that is to say, the possible "madadaptive responses in those places that are stigmatized as being inherently more prone to climate-induced violence" (Adams et al. 2018, p. 200).

### 6.2.2 Implications for Organizational Research and Research on IOs

In addition to security theory, the thesis relies on organizational research and thereby incorporates various theoretical strands of organizational research in an all-encompassing way. The aim is to find out which theoretical 'lens'—or, rather, which combination of theoretical 'lenses'—offers the most explanatory power for the empirical puzzle at hand.

#### 6.2.2.1 Implications for Leadership Research

First, the thesis contributes to the organizational research that centers on the role of leadership in IOs. A recent comparative study by Hall and Woods examines the effectiveness of IO leadership, and the UNHCR is one of the 11 IOs under consideration (Hall et al. 2015). It shows that the UNHCR is an example of an IO that follows a comparatively transparent approach when it comes to finances, expenses and the election procedures of the High Commissioner. Even though this analysis falls into line with the broad arguments of the study by Hall and Woods, it challenges another argument of organizational research.

Concerning the leadership at the UNHCR, organizational research often refers to the 'moral prestige' of the top management and the interplay between moral authority and leadership skills (Loescher 2014, pp. 215–217). This thesis gives flesh to these attributions to the High Commissioner. The analysis shows that the aspect of 'moral reputation' is not significant in explaining, first, whether the leadership is effective within the UNHCR and, second, the real-life impact of the leadership in the field of concern. By discussing the often-cited moral reputation of the UNHCR, it shows that Guterres did—to some extent—leverage the moral standing of his organization. His speeches in front of the UN Security Council, at international forums and in the media were uncomfortable for many states, especially in the global North. The High Commissioner called states to take responsibility, and, by means of his wordings, he left the comfort zone of his organization. Therefore, the way in which Guterres interpreted his role as High Commissioner and what he thought the UNHCR's future workload should be were relevant. Nevertheless, concerning Guterres and his EDP engagement, the 'moral standing' did not turn the balance. Even though Guterres could profit from the UNHCR's reputation in the global realm, this 'moral standing' did not automatically result in the leadership's strong involvement in the topic of EDPs. In a similar vein, the organization's standing as the leading refugee agency does not explain its success in becoming a Standing Invitee to the Nansen Initiative and a respected organization on the Protection Platform.

In other words, the UNHCR approached the topic of EDPs because the then High Commissioner António Guterres had put the topic on his agenda. Therefore, concerning the theoretical 'lenses' of the research on the 'character of the bureaucracy' some of the assumptions have proven to be relevant. First, the research on the leadership in IOs (cf. Sect. 3.3.3) and the respective theoretically deduced assumptions have suited the empirical puzzle and thereby helped to explain the UNHCR's approach. Second, the different perspectives on an IO's workforce and staff were not decisive in explaining the empirical puzzle. The small group of individuals at the UNHCR who pushed the topic were not able to provoke an overall organizational change or the opening of the UNHCR toward the topic. Thereby, this thesis adds to the organizational theory and argues that the bureaucracy mattered less than expected.

#### 6.2.2.2 Implications for Research on Organizational Change and Learning

Second, the thesis adds to the organizational research that considers organizational change and organizational learning. Here, it takes up the definition of 'learning' by Haas: "The ultimate purpose is redefined, as means as well as ends are questioned [and the] behavior changes as actors question the original implicit theories underlying programs and examine their original values" (Haas 1990, p. 3). Based on this understanding of 'learning', this organizational analysis does not identify organizational learning throughout the whole period of analysis. Instead, the UNHCR adapted its behavior and instigated new activities but without questioning its underlying values (the Refugee Convention). Haas describes 'adaptation' as "the ultimate purpose of the organization is not questioned [. . .], the emphasis is on altering means of action, not ends" (Haas 1990, p. 3). Following this understanding of 'adaptation', Chap. 5 establishes that, from 2011 onwards, the UNHCR went through a phase of adaptation. The analysis falls into line with previous organizational research that establishes that "IOs with a deeply rooted bureaucratic culture are difficult to change" (Nielson et al. 2006, pp. 108–109) and with research that looks into the different bureaucratic styles (Enkler and Grohs 2017, p. 255). More specifically, the thesis connects to the research on organizational change that considers different factors, such as the increasing heterogeneity of member states, the evolution of UNHCR communication structures and UNHCR formats for more transparency (Dingwerth et al. 2014, p. 203).

Regarding the thesis' contribution to research on organizational change, the findings respond to the call of Dingwerth et al. (2014) to examine the triggers that lead to organizational change. Dingwerth et al. demand that researchers pay more attention to the question of whether IOs respond to their wider target group in an ad hoc mode or in a strategic manner (Dingwerth et al. 2014, p. 207). Concerning the period from 2005 until 2015, the analysis at hand contributes to answering this question. The analysis shows that the UNHCR coped with the triggers of human mobility strategically: the early statements by Guterres are in line with his goal that the UNHCR would stay relevant in the international realm and that it would remain the leading international actor on all matters of human displacement and mobility. The intra-organizational development throughout the 10-year-long period of analysis shows that the organization intended to position itself strategically in the international forums on the topic—despite the setback due to the states' responses in 2011. Thereby, this analysis opposes the research on the UNHCR's climate change approach, which argues that the organization's interest in the topic was *not* motivated by relevance (Hall 2015, p. 93).

In other words, the UNHCR approached the topic of EDPs first to remain relevant in the context of an ever-increasing 'humanitarian marketplace' and second regarding the new international developments in the context of human mobility, that is, a future rise in human mobility in the context of environmental changes. Hence, the theoretical 'lenses' of the research on the 'organizational environment' (including cooperation and competition between IOs) and the research on 'organizational learning' have proven to be relevant to and suitable for the empirical puzzle at hand (cf. Sect. 3.3.4).

#### 6.2.2.3 Implications for Research on the Autonomy of IOs

Third, the thesis adds to the research on the autonomy of IOs, because it questions the role of states in the UNHCR's EDP approach. Hall establishes that "member state preferences alone cannot explain the UNHCR's approach to climate change and displacement" (Hall 2013, p. 104). This thesis concurs with Hall's argument, because it shows that states did not push for an opening toward the topic of EDPs. Instead, the thesis depicts an IO that acts independently and autonomously (despite it being highly dependent on the financial contributions of states). More specifically, the UNHCR's involvement in the Nansen Initiative can be traced back to the efforts of the committed individuals at the UNHCR; that is, it was the UNHCR and not the Nansen states that reached out to the initiative. The Nansen states were careful after the negative responses from states at the UNHCR Ministerial Meeting in 2011. The developments in the organizational environment influenced the UNHCR's interest in EDPs, but no external actor explicitly delegated the organization to the topic.

In addition, the thesis contradicts some existing theoretical assumptions from organizational theory concerning the mandate of IOs. The thesis establishes that, in the EDP context, the UNHCR's mandate was not the leading explanatory factor. This observation concurs with Siebenhüner, who claims that "it is not necessarily the formal mandate that explains the degree and type of influence of IOs" (Siebenhüner and Biermann 2009, pp. 265–266). Previous studies on the UNHCR's mandate, for example Loescher and Milner (2011, p. 199), shed light on the organization's workings until the early 2000s. The factors that become apparent in those earlier studies, such as the effect of the UNHCR's organizational structure and the role of donor states, also guide this thesis' approach. Concerning the topic of EDPs, it shows that a well-established IO, such as the UNHCR, still runs the risk of juggling its organizational responsibilities on the one hand and its organizational interest on the other hand.

The theoretical 'lenses' of those theories that look inside an IO are necessary to explain the empirical puzzle. With the aid of the perspectives of the intra-organizational theories (cf. Sect. 3.3.4), the analysis shows that the UNHCR used its organizational norm of 'protection' from 2011 onwards to advocate its role in EDP protection. In addition, the organization played a part in the 'climate refugee' debate because it felt obliged to protect its mandate—the Refugee Convention. Even though the UNHCR's autonomy was apparent in the period of analysis (e.g. the autonomy from states, cf. Sect. 5.2.2), the organization overall did not use its autonomy. The question of why the UNHCR did not succeed in institutionalizing the topic until 2015 is beyond the research scope of this analysis. The analysis nevertheless reveals that the UNHCR showed autonomous behavior, because the topic arose from the organization itself and the member states did not bring the topic to the UNHCR. Yet, this seemingly self-determined behavior was limited to the public speeches of Guterres and some involved individuals at the UNHCR. In other words, the Ministerial Meeting of 2011 exemplified that the organization has little leeway if its projects do not fall under earmarked funding and if its own members of staff—in the majority—do not support the new approach. In this context, the thesis falls in line with recent research that discusses how administrative self-constraint sometimes is in the best bureaucratic self-interest because, by backing down, IOs avoid the risks of a proactive behavior (Bauer et al. 2017, p. 190). Thereby, the thesis concurs with recent research, which also points out that the topic is not yet part of the organizational culture (Goodwin-Gill and McAdam 2017, p. 29) due to the resistance from states and staff (Hall 2013, p. 105). Despite these challenges, it is remarkable that the UNHCR underwent the development as sketched in this thesis from 2005 until 2015.

#### 6.2.2.4 Implications for the Research on IOs

Over the last decennia, IR scholars have taken the first steps toward organizational theories to explain the role of IOs in global governance and in inter-state relations. The IR community overall perceives this endeavor, that is, the theoretical opening and mutual stimulation of IR and organizational theories, as fruitful (Dingwerth et al. 2009, pp. 25–26). Nevertheless, despite this initial approximation, a research gap still exists when it comes to including organizational theory in IR research, and the thesis at hand takes one step toward filling this gap.

The thesis thus gives testimony that the research on IOs contains important policy implications. The work at hand builds on earlier research on the behavior of IOs, and it responds to the call of scholars for more research on the interplay of an IO's secretariat, its executive organs and its member states (da Conceição-Heldt et al. 2014, p. 5). Hence, the thesis adds to the research on the UNHCR that focuses on the question of why the UNHCR's debate shifted in the 2000s and 2010s. Hammerstad argues that "international protection returned to center stage of its [the UNHCR's] official discourse" (Hammerstad 2014, p. 159) because the organization wanted to set itself apart from the other actors in the international realm. The analysis at hand follows- up on Hammerstad's hypothesis, and the thesis indicates competition between the IOM and the UNHCR on the question of which organization is the leading actor on the topic of EDPs. The public image of both organizations is that they extensively collaborated before the COP in 2009 and in the Nansen Initiative until 2015. Both organizations also declared their common success in bringing the topic of EDPs into the Cancun Declaration of 2010. However, this thesis also portrays the competing character of the two organizations, and the analysis outlines the battle over financial resources, international standing and expertise on the topic. Thereby, the thesis reflects and complements earlier UNHCR research.

By this means, this thesis adds to the research on IOs because it describes the importance of considering the organizational environment. The thesis concurs with the recent assessment by McAdam (2017) that, until 2015, the UNHCR's engagement in climate change and disaster-related displacement predominantly focused on the organizational environment. To put it differently, a relationship exists between the UNHCR's interest in EDPs and the events of the environment, for example the Cancun Declaration in 2010, the disappointing ExCom meeting in 2011 and the Nansen Initiative. Following McAdam, these events were "a direct response to a concerted effort by UNHCR during 2011 to encourage states to work toward the formulation of a 'global guiding framework' on displacement relating to climate change and natural disasters" (McAdam 2017, p. 1521). As an example, Marine Franck of the UNHCR explained that the UNHCR intended to bring human mobility into the Paris Agreement (COP 21) to address the action gap with solid tools whilst being able to guide states' work on the topic (Franck 2015, p. 11). The analysis showed that the UNHCR's cooperation as well as its competition with the various other international actors in the field of migration and displacement (e.g. the IOM, the IDMC, the NRC, etc.) had a substantial impact on the way in which the organization positioned itself on the topic of EDPs (see Sect. 5.2.1). For the research on IOs, this finding implies that in-depth organizational studies should not analyze IOs in isolation; instead, the finding of this thesis is that IOs are always caught between their internal organizational goals, strategies and actions and the external environment, including other international actors and international politics. This finding is one explanation for why the UNHCR approached the topic of EDPs, and the thesis hereby falls into line with the research on IOs that concludes that the "UNHCR's early work on climate change, disasters, and displacement was visionary, if premature, in creating a global roadmap for action" (Goodwin-Gill and McAdam 2017, p. 23).

### 6.2.3 *Implications for Refugee Research and IR Research*

To start with, this thesis questions the previous results of refugee research. Some scholars argue that the UNHCR's role in the EDP debate was inevitable due to the 'climate refugee' references (McAdam 2014, p. 13). This analysis shows that even though the refugee references by politicians, the media and NGOs from 2005 until 2011 triggered a response by the UNHCR, the 'climate refugee' depiction does not suffice to explain the organization's active role in—among others—the Nansen Initiative after 2011. Instead, the thesis outlines the various aspects that factor into the UNHCR's approach to EDPs, for example the personal interest of the High Commissioner and the effort of the small group of UNHCR staff. However, the thesis also adds to human mobility research in general and refugee research in particular. The research question of this thesis lies at the heart of one of the biggest humanitarian challenges of the twenty-first century, namely disaster displacement. The aim of this thesis was to avoid falling into the trap of discussing the non-applicability of the Refugee Convention in the context of EDPs once again. Instead, the different security references sensitized the analytical focus on the affected individuals. The research design of this thesis therefore intended to shed light on the way in which the UNHCR, one of the leading international actors on the topic of human mobility, has begun to develop an interest in disaster displacement. In this context, the UNHCR's picking up of the topic merely represents the beginning. The future will most likely show more examples of organizations that feel the impacts of environmental changes on their daily work. The IDMC, the IOM and the NRC and their respective publications on and involvement in the topic of EDPs thereby constitute the first examples of this development. Hence, the organizational analysis of this thesis is a precursor of more academic attention to the topic of IOs and their view of disaster displacement. Therefore, some follow-up research questions can be raised: First, how will the impacts of environmental changes challenge the actors of the international community? Second, how will these actors respond to the challenges? Moreover, will they thereupon adapt their respective mandates? Third, to what extent will this development have an impact on the protection of those who fall under the Refugee Convention?

By this means, refugee research has to consider the changing nature of displacement at the start of the twenty-first century—although the affected persons are not all refugees. Therefore, the research question of this thesis touches on the asylum–migration nexus. This implies that refugee research cannot ignore the other forms of human mobility, such as migration and planned relocation. In reality, the routes used by migrants and asylum seekers (with a prospective refugee status) overlap, and it would therefore not do justice to refugee research to draw artificial boundaries between these different forms of human mobility.

Furthermore, the thesis at hand adds to IR research in a number of ways. First, it concurs with the argument that IR research can profit from organizational research (Dingwerth et al. 2009, pp. 25–26; Ege and Bauer 2013). If IR research intends to include the behavior of IOs in international politics, it needs to dig deeper than the IOs' official agenda and façade; instead, "it is the bureaucracies within these international organizations—their staff and leaders and the way they structure their work—that matter" (Biermann and Siebenhüner 2009a, pp. 345–346). Therefore, those areas of research that have a stronger focus on the intra-organizational events should add to the theoretical 'lenses' of IR research. Second, the thesis shows that IOs are not merely the hand puppets of states. The analysis falls into line with the research that questions this theoretical assumption of principal–agent theory (Siebenhüner and Biermann 2009, pp. 265–266). The analysis of the UNHCR's EDP approach illustrated the self-determined nature of the organization's approach, and the thesis points out that states were in fact not the ones that pushed the UNHCR toward the topic. These observations add to IR research because they shed additional light on the way in which IOs interact with states. In addition, organizational research helps to clarify the means by which IOs lobby for their cause. In the case of the UNHCR, the thesis outlines how the organization has to campaign for the global refugee regime while simultaneously keeping the states' interests in mind. Similarly, Betts calls for the study of forced migration to become part of IR research because it has a "potentially wide-ranging theoretical contribution to make to the discipline" (Betts 2009, p. 2) due to its focus on globalization, international cooperation, security and North–South relations.

### 6.2.4 *Critical Reflection*

First and foremost, it has been a challenge to approach the research interest of this thesis, the UNHCR. The analysis shows that, as an outsider, it takes a lot of effort to look behind the façade of the UNHCR, and it is difficult to uncover the steps taken by the organization toward the topic of EDPs. The thesis touches a sore spot of the UNHCR—its accountability as opposed to its organizational environment. If an IO's agenda setting is not transparent and comprehensible to the outside world, the organization is in danger of losing its reputation, because it is difficult for its partner IOs and its member states to predict its next steps. The UNHCR realized this weak point in its decision-making on EDPs and learned from the negative experiences during the Ministerial Meeting in 2011. As a result, the organization introduced the Global Management Accountability Framework and the Results Framework and started a Global Learning Center for greater coherence of activities and programs (Türk and Eyster 2010, p. 168). In addition, the UNHCR took a backseat in the Nansen Initiative to avoid stepping on the states' toes once again. However, during the research process, it did not prove to be easy to uncover this window-dressing and strategic behavior of the organization in regard to its EDP approach.

By reflecting on the choice of theory, it is apparent that the chosen approach, to include a broad range of theoretical 'lenses' on the UNHCR's EDP approach, has its pros and cons. The thesis examines the UNHCR from a variety of (theoretical) angles, minimizing the likelihood of missing an explanatory factor. The intention was to put the events around the organization's EDP approach at the center of attention and—instead of theory building—to use the organizational theories as a means to reconstruct and understand the UNHCR's steps toward the EDP topic. Next to this, the organizational theories proposed different arguments for the empirical puzzle at hand. Nevertheless, the testing of theories will not generate one 'winner' among them. The conceptual design can merely point out which theoretical 'lens' has been helpful in uncovering a specific aspect of the UNHCR's approach to EDPs.

Nonetheless, a critical reflection on the choice of theory must include a word or two about the theories selected for the analysis. To start with, a case study design is not able to 'prove' or 'falsify' a theory in the strict sense of the words (Blatter et al. 2007, p. 186). For the empirical puzzle at hand, a combination of theories with an intra-organizational and an external perspective proved to be most suitable: the 'member states' did not push the UNHCR's EDP approach by themselves, but the broader organizational environment played a large role in bringing the topic into the organization's view. Additionally, the theoretical 'lenses' that take the leadership into consideration were helpful in reconstructing the influence of High Commissioner Guterres. On a more critical note, and due to its strong focus on the empirical puzzle, the thesis does not manage to contribute to theory building. Instead, it uses the theoretical background of the conceptual design as a 'toolbox' that helps it to grasp the empirical puzzle. By this means, the thesis avoids falling into the traps of, first, weighing the theories against each other and, second, oversimplifying the theories to reach the intended goal.

Another challenge in the research design was the data collection, more specifically the qualitative expert interviews. Prior to conducting the field research in Geneva, it was not apparent how many individuals would fulfill the criteria of an 'expert' (see Sect. 4.3.1). In Geneva, the challenge was to ask only those individuals who could contribute substantially. In the end, the circle of experts was straightforward and the challenge entailed presenting the findings of the analysis while simultaneously guaranteeing the anonymity of the experts. Moreover, the expert interviews took place when the follow-up of the Nansen Initiative—the Platform on Disaster Displacement—was already up and running. Nonetheless, the research interest of this thesis lies in the period when the Nansen Initiative had just been a fixed idea of a small group of states. During the interviews, it showed that, when the experts recapitulated the early 2010s, their accounts were overshadowed by the later success of the Initiative. In addition to this, some of the expert interviews were challenging to conduct because the experts had to talk about their current employer—sometimes at their current workplace. In the first phase of the interview, some experts thus tended to present a perfect image of their employer; fortunately, this precautionary behavior declined throughout the interviews. In retrospect, participatory observation over a longer time span would have given even more insight into the workings of the UNHCR. Unfortunately, the work on this thesis only started at the end of Guterres' reign; therefore, the method of participatory observation was no reasonable option. In addition, a questionnaire did not seem to be a sensible method of data collection because the experts might feel even more hesitant about commenting on their current employer in written form. Furthermore, some of the more critical interview questions are best asked face to face. Depending on the interview situation, the interviewer can adapt the wording and the phrasing of the questions, and thereby expert interviews represented the most likely means to receive an answer. Lastly, at the beginning of the interviews, some of the academic experts were very keen to discuss the concrete research question of the thesis, including the theoretical background. Here, the challenge was to provide an abstract summary of the research design that did not interfere with the subsequent interview. Overall, the choice of experts has a direct impact on the information gained, which leads to the conclusion that, in the case that the analysis is repeated in a few years—including a different set of questions and interviewees—the possible outcome will differ from the one gained in the interviews at hand.

Concerning the research design of this thesis, it proved helpful to have different methods that complement each other regarding the data collection. The expert interviews succeeded in adding to the literature review of the primary and secondary data. The interviews introduced background information and personal accounts of the events from 2005 until 2015. In addition, it proved to be essential to talk to the people who worked at the UNHCR or who worked with the UNHCR staff during the period of analysis. Nevertheless, the literature review of the primary and secondary data was indispensable and represented the core of the analysis. From a critical stance, the field research stressed the importance of the intra-organizational relations. That said, the field research triggered an interest in a comparative case study, such as the IOM, to add flesh to the overall findings of this thesis.

## 6.3 Outlook

*There are no humanitarian solutions to humanitarian problems, only political ones. But our strictly humanitarian character facilitates our work.* (Guterres 2012a, p. xx)

The discussion of the findings point out a few open questions that remain to be addressed by follow-up research. Recent research tackles aspects of EDPs that scholars have previously bypassed. Examples are the research that centers on the visualization of EDPs by means of visual discourse analysis and the research that poses questions regarding the politicization or depoliticization of global warming (Methmann 2014). Another example is the research on the climate security discourse's evolvement from conflict to resilience (Boas and Rothe 2016). Migration research has also recently started to focus on 'migration management', understood as a policy discourse that counters security-oriented immigration policies (Scheel and Ratfisch 2014, p. 925). The scientific debate on the topic of climate change and migration points to the politics of power that are hidden behind the leading 'climate refugee' debates. Baldwin and Bettini (2017) critically discuss the epistemic community of researchers and experts that has formed on the topic. They argue that this epistemic community follows a shared set of assumptions on the topic of climate change-induced human mobility (Baldwin and Bettini 2017, p. 5).

Nevertheless, this thesis concurs with Siebenhüner who argues that "more research on the role of IOs in global environmental governance is needed" (Siebenhüner and Biermann 2009, pp. 268–269). The thesis at hand investigates organizational learning and organizational change at the UNHCR, but it remains to be seen whether the UNHCR's development on the topic of EDPs will be substantial and of long duration. Even though this thesis calls for more case studies on different IOs (see Sect. 6.2.4), this scholarly effort will not suffice. Instead, the focus should simultaneously rest on the question of the kind of impacts or outcomes that this organizational behavior entails. In other words, if organizational learning or organizational change occurs at the UNHCR in the future, what will the implications be for refugee protection and IDPs? That said, organizational research should not restrict itself to identifying a certain organizational behavior, but it should also examine the effects of an IO's behavior on its persons of concern, for example refugees. Taking up the research question of this thesis, subsequent research should keep track of whether the UNHCR succeeds in opening its workload to EDPs and what this step means for its mandated responsibilities.

Therefore, a need still exists for more academic research on the questions of, first, where the environmentally displaced will go and, second, under what conditions they will stay or return. The thesis accords with McAdam, who states that the reason for more research on EDPs is that political responses to the topic should be "guided by considered, well-informed research, not by sensationalism, assumptions or fear" (McAdam 2010, p. 4).

## *6.3.1 Policy Implications*

### 6.3.1.1 Institutionalizing a New Topic

The first policy implication of this thesis is that the UNHCR will need to mainstream the topic in the organization if it intends to keep working on EDPs. The analysis of this thesis shows that a variety of opinions from different actors exist on the question of how the UNHCR should approach the issue of EDPs. In other words, the High Commissioner, the environmental officers and the overall staff at the UNHCR all held different views on what the organization's position on the topic should be. In addition, the states' and partner IOs' perspectives on the UNHCR's approach differed immensely. In this context, it is helpful to remind oneself of Janmyr's three levels of institutional practice: the practice of the UNHCR's political organ, the practice of the upper bureaucratic levels and the actual conduct of the organization (Janmyr 2016, p. 52). This thesis is supportive of the idea that organizational practice is the outcome of a negotiation among multiple levels of an IO. The UNHCR's hidden approximation to EDPs lacked intra-organizational mainstreaming and was not helpful in positioning the organization among the other international actors. In the first step, the UNHCR should decide once and for all whether its goal is to add more people of concern to its workload. The new High Commissioner, Filippo Grandi, and the upper bureaucratic level need to carry out a fact-finding mission throughout the organization and ultimately decide on the position of the organization. By means of this clear positioning, the leadership has to inform all the staff members about the decision. In the case that the UNHCR decides not to approach the topic of EDPs, it should declare this position openly and act in concert on this decision.

### 6.3.1.2 Responding to a Changing International Environment

The second policy implication of this thesis is that the case study describes one means of adapting to a changing organizational environment. IOs are involved in almost all problem areas in the international sphere to carry out programs or implement rules (da Conceição-Heldt et al. 2014, p. 8). The presence of IOs is immense, but, due to their different means of exercising their influence, their contributions to decision-making are often overlooked. In the words of Abbott, "IOs provide an important supplement to decentralized cooperation that affects the nature and performance of the international system" (Abbott and Snidal 1998, p. 29). Hence, the UNHCR's approach to EDPs might be used as a blueprint by other IOs. The thesis shows how an IO upholds its normative agenda, that is, its mandated role in international protection, in times of new challenges and a changing international environment. Ultimately, as of 2018, the UNHCR did not fully incorporate the topic of EDPs into its workload, and, even though this might be a backlash to former High Commissioner Guterres, it nevertheless shows the resilience of organizations to the ever-changing nature of international relations. Thereby, the case study of the

UNHCR's EDP approach provides lessons for other IOs on the question of how to guard one's institutional framework (Loescher et al. 2012, pp. 129–130). The thesis shows that the UNHCR had to juggle, first, the possibility of losing its international standing on matters of protection and, second, the protection of the Refugee Convention and its responsibilities to refugees. Nevertheless, in the long run, the organization will have to decide on a response to the new developments in human mobility to avoid making itself irrelevant and giving the impression of being an antiquated organization that closes its eyes to the real world. Thereby, the future will show whether the resolutions of the UNHCR on EDPs were a wise step or whether the UNHCR should have hopped on board and adapted to the new reality much faster.

#### 6.3.1.3 Emancipating Itself from States' Interests

The third policy implication of this thesis regards the politicization of refugees. Throughout the existence of the refugee regime, state interests have interfered with the work of the UNHCR. In this context, the refugee agenda must compete with other topics—for example security-related, economic or environmental issues—that also require the attention of states (Loescher and Milner 2011, p. 190). The UNHCR's work has always been highly influenced by politics, even though the states originally intended to establish an international body that is separate from power politics and that follows its strictly mandated goals. This thesis carves out the way in which national interests influenced the UNHCR, for example by the restraints due to the financial contribution of states. The period of analysis from 2005 until 2015 represents an episode during which the UNHCR was led by a former politician, António Guterres, who has considerable national and international experience. From the beginning of his reign, Guterres jumped into the political debates and adopted a stance on the matters that he presumed to be relevant to the UNHCR's role. In this context, the thesis shows a shift from earlier High Commissioners toward Guterres: the former fought for a role of the UNHCR in terms of having enough funding and being respected among other UN organs, and the latter fought for a role of the UNHCR in international politics. The result of Guterres' 'surge ahead' was a falling-out with many states in 2011. In the debates on EDPs as well as in the Nansen Initiative, his decisions forced the UNHCR to take a back seat. That said, even though the UNHCR's leadership—that is, the High Commissioner—was keen to discuss an organizational opening, this "has not necessarily served the fulfillment of UNHCR's core mandate or furthered the interests of refugees" (Betts et al. 2012, p. 162).[1] Despite these challenges, the High Commissioner contributed to general awareness raising of the topics of EDPs, disaster displacement and climate change.

[1]Nevertheless, it was not only the High Commissioner whose role was under the influence of politics. Other organizational analyses show that the ExCom has become more politicized over time to the extent that some scholars even argue that "it is no longer an effective decision-making body" (Loescher 2014, p. 221).

#### 6.3.1.4 Bringing the People into the Conventions

The fourth policy implication is that the UNHCR brought the people of concern to the center stage. By this means, this thesis adds to the corpus of literature (among others, da Conceição-Heldt et al. 2014, pp. 4–5) that sheds light on the consequences of IOs' actions for international relations and global politics. In the context of global governance, analyses often overlook and underrate the influences of IOs, and this neglect will "mislead conclusions about the state of the world politics, and to result in an overemphasis on state power" (Biermann and Siebenhüner 2009b, p. 2). In this context, Biermann and Siebenhüner outline the problem that the policy outcomes that were influenced by IOs "are thus likely to be overlooked" (Biermann and Siebenhüner 2009b, p. 2). As regards the wording, Gemenne (2015) recommends talking about 'climate refugees' when referring to EDPs. He argues that this wording "recognizes that these migrations are first and foremost the result of a persecution that we are inflicting on the most vulnerable" (Gemenne 2015, p. 71). The aspect of responsibility becomes central in this context, and the 'human' aspect of 'human mobility' takes center stage. Applying Gemenne's proposition to this thesis, it shows that there still is much to be undertaken before states will see EDPs as part of their genuine responsibility. However, the thesis exemplifies how one of the leading international organizations—the UNHCR—has become involved in the topic. It thereby indicates how the UNHCR has ultimately followed the same agenda as described by Gemenne: the UNHCR's goal was to put EDPs on the agenda of international forums as well as on the policy agenda of states. Therefore, the organization campaigned for a human perspective, not a security perspective, on the topic.

Hence, the fourth policy implication is that the UNHCR has shown one possible way to lobby for EDPs without losing sight of one's own organizational agenda and mandate. The UNHCR experienced an evolution regarding how to talk about EDPs and eventually opted for a human security-centered depiction. This decision—made alongside other NGOs and IOs—also influences the way in which the international community talks about EDPs today. Global environmental governance usually has a relatively low domestic priority in many states. The reason is that environmental policies often do not produce immediate benefit for governments, which makes it a challenging field of policy (Siebenhüner and Biermann 2009, pp. 265–266). In addition, IOs usually only receive limited support from governments. In the policy field of EDPs, it took most states quite a long time to realize the relevance of the topic, and the IOs such, as the UNHCR and the IOM, therefore had a large role in influencing the way in which states perceived the issue and how they talked about it. The success of the IOs' advocacy on the topic becomes clear in the international agreements that states adopted during or shortly after the period of analysis, for example the Cancun Agreement, the Nansen Protection Agenda, the Sustainable Developments Goals and the Paris Agreement.

### 6.3.2 Way Ahead

When trying to look forward, that is, into the future of the refugee regime and human mobility, the forecast needs to take the following arguments into account.

First, the refugee regime is specific in the sense that "states would almost certainly not have agreed to [the regime] at any subsequent historical juncture" (Loescher et al. 2012, pp. 128–129). In other words, the Refugee Convention is a product of its time and is closely linked to the post- World War II period. In a 2016 article, Collier critically argues that the way in which the UNHCR deals with refugees is still stuck in the year 1951 by assuming that refugees merely need food and shelter (Collier 2016). In 1951, the global expansion of the Refugee Convention could not have been anticipated, and the additions to the workload of the UNHCR, for example IDPs, was not part of the original organizational set-up. Given that the organization has kept its original mandate throughout the ups and downs of global politics since 1951, an imminent change in the refugee regime is rather unlikely.

Second, it is apparent that "global challenges such as climate change [and] migration [. . .] can be addressed only through the joint effort of a range of actors, both public and private" (Steffek and Kortendiek 2018, p. 204). Even though, at the end of the last century, Europe "became more peaceful and less 'promising' in terms of the 'problems' the IOs [IOM and UNHCR] are tasked to address" (Pécoud 2015, p. 20), this thesis shows that—as of 2018—this statement is no longer true. In other words, the UNHCR and the IOM no longer suffer from a declining workload. Instead, the thesis sheds light on the challenges of human mobility in the twenty-first century and a changing international environment, in which the UNHCR is merely one actor among many others. Irrespective of the diverse factors that lead to human mobility, the post-2015 period has witnessed a rise in attention toward people on the move. Whereas this thesis mainly considered an IO's global perspective on EDPs, a stark contrast can be seen regarding the national sentiments of the 'receiving countries'. In 2015, the diverse reactions of European countries to people who fled to Europe exemplify the political nature of the topic. Compared with other world regions, the number of people who seek to take refuge in European countries is still small. However, the year 2015 has already made waves in terms of border restrictions, the building of fences and limitations on asylum seekers. Discussions about policies that address the root causes of migration and displacement are on the rise, and the integration policies, the EU–Turkey refugee agreement, the future organization of FRONTEX, and the repatriation of asylum seekers are just a few of the much-discussed topics in this context. On the one hand, 2015 might signify a chance for the UNHCR. The European states' acknowledgment that the topic of human mobility is their concern hereby represents an opportunity for the UNHCR: if more states decide to focus on the countries of origin and on the root causes of human mobility in general, there is a chance of more financial assistance for the UNHCR. On the other hand, the European states' reactions since 2015 represent a challenge for the UNHCR, because the organization must repeatedly fight for the application of the Refugee Convention, and it must admonish states for their restrictive immigration policies.

Third, learning from the findings of this thesis, there is no likely scenario that would lead to states' putting down the UNHCR and—as a result—assigning all aspects of international protection to themselves. In reality, states' behavior forecasts the opposite direction: contentment that it is best if the straining topic of human mobility is organized and implemented by an IO in return for a small financial contribution of states. The current state of international politics makes a future scenario in which states will share the UNHCR's workload among themselves less likely. Instead, states' policies that further the national interests are increasing, and some experts on the topic even recognize a "shift from asylum to containment where Western states have largely limited the asylum they offer to refugees and have focused on efforts to contain refugees in their region of origin" (Loescher 2014, p. 219). Here, it will become important whether the international community conceives human mobility as a security issue, a refugee issue, a human rights issue, a migration issue, an environmental issue or a humanitarian issue (McAdam 2011, p. 157). The international developments on the topic that took place from 2005 until 2015 are remarkable, and a community of committed states, IOs and NGOs emerged to pick up the topic of EDPs. Many participants in the Nansen Initiative emphasized that the UNHCR's expertise and involvement helped considerably in their common endeavor. Nevertheless, by 2015, the Nansen Initiative actors had not fully achieved their goal "to develop a guiding framework or instrument for the protection of people displaced externally due to sudden-onset natural disasters, including those related to climate change" (Nansen Initiative 2011, p. 19). Instead, the Nansen Initiative's follow-up, the Platform on Disaster Displacement, took over in 2016 and has actively worked toward more states acceding to the Nansen Protection Agenda. On a positive note, this platform responds to scholars' wishes and calls (among others, Loescher 2014, p. 220) for more awareness on the topic in the international community. During the Nansen Conference in 2015, a large number of states endorsed the Nansen Protection Agenda, which shows that the topic of EDPs is recognized as being one that needs states' attention. Additionally, it reflects the national interest of states in involving themselves, because only if they take part in the process of establishing a new platform can they guarantee their influence.

Fourth, the forecasts concerning the future number of EDPs need to be analyzed carefully, especially concerning the methodology of the research. It will be the role of scholars to add empirical facts to the heated debates in the media and among decision-makers. Thereby, academia needs to rise to the occasion of the important role that it plays regarding research on human mobility in general and on EDPs in particular. Nonetheless, the thesis also discusses the fact that there usually are diverse factors that lead to human mobility; that is, it is difficult to separate environmental triggers from other causes of displacement and migration. In other words, "there are no criteria to determine where to draw the line between voluntary movement and forced displacement" (Kälin 2010, p. 90). Academic research will have to face this challenge by means of sensible research, which does not become embroiled in the highly political nature of the matter. In this regard, it is important to bear in mind that disaster displacement is one of the biggest humanitarian challenges of the twenty-first century (McAdam 2016). Hereby, it is highly relevant to differ and to make clear that "many hydro-meteorological disasters triggering displacement would occur regardless of climate change, and even when they are linked to climate change, such causality is difficult, if not impossible, to prove in a specific case. Furthermore, disasters such as volcanoes or earthquakes with similar impacts on people and on their movements are not linked to climate or to climate change, but there is no reason why people displaced by those events should be treated any differently from those affected by climate impacts" (Kälin 2010, p. 85).

Fifth, the future will show how the refugee protection regime will develop and whether the Refugee Convention will be adapted and rejuvenated. Furthermore, the role of the UNHCR in the context of human mobility is likely to have to adjust further to the new drivers of movement. Scheel and Ratfisch describe the deepening and widening of the UNHCR's role in the global protection regime (Scheel and Ratfisch 2014, p. 938), but this stretching of the UNHCR's workload also has its limitations: the UNHCR will not be able to overstretch infinitely without losing sight of its original 'people of concern'. In the case that the UNHCR intends to widen its area of involvement, it would need the support of states to change its statute. This widening would automatically involve careful consideration of the UNHCR's budget, and there is a strong chance that the organization would ultimately end up with less—in terms of influence, financial contributions and reputation—than today. Another possible development constitutes a deepening of the organization's outreach. This would imply that the UNHCR centers its attention on its core mandate and thereby builds its reputation as the world's leading refugee organization. Based on the analysis of this thesis, deepening seems to be the less likely future scenario. The underlying reason is that this thesis shows that IOs are unable to act in isolation and that they always are a reflection of the respective period of time. In the context of the UNHCR, the issue of human mobility is not static and predictable; instead, the factors that lead to flight are overlapping and permanently changing and will continue to be so in the future.

## References


Abbott KW, Snidal D (1998) Why states act through formal international organizations. J Confl Resolut 42:3–32

Adams C, Ide T, Barnett J, Detges A (2018) Sampling bias in climate-conflict research. Nat Clim Chang 8:200–2005

Baldwin A, Bettini G (2017) Introduction: Life adrift. In: Baldwin A, Bettini G (eds) Life adrift: climate change, migration, critique. Rowman & Littlefield, Washington, DC, pp 1–22

Baldwin A, Methmann C, Rothe D (2014) Securitizing 'climate refugees': the futurology of climate-induced migration. Crit Stud Secur 2:121–130. https://doi.org/10.1080/21624887.2014.943570

Bauer MW, Knill C, Eckhard S (2017) International public administrations: a new type of bureaucracy? Lessons and challenges for public administration research. In: Bauer MW, Knill C, Eckhard S (eds) International bureaucracy: challenges and lessons for public administration research. Palgrave Macmillan, London, pp 179–198

Betts A (2009) Forced migration and global politics. Wiley-Blackwell, West Sussex, UK

Betts A, Loescher G, Milner J (eds) (2012) UNHCR: the politics and practice of refugee protection, 2nd edn. Routledge, London

Biermann F, Siebenhüner B (2009a) The influence of international bureaucracies in world politics: findings from the MANUS research program. In: Biermann F, Siebenhüner B (eds) Managers of global change: the influence of international environmental bureaucracies. MIT Press, Cambridge, MA, pp 319–350

Biermann F, Siebenhüner B (2009b) The role and relevance of international bureaucracies: setting the stage. In: Biermann F, Siebenhüner B (eds) Managers of global change: the influence of international environmental bureaucracies. MIT Press, Cambridge, MA, pp 1–14

Blatter JK, Janning F, Wagemann C (2007) Qualitative Politikanalyse: Eine Einführung in Forschungsansätze und Methoden. VS Verlag für Sozialwissenschaften, Wiesbaden

Boas I, Rothe D (2016) From conflict to resilience? Explaining recent changes in climate security discourse and practice. Environ Polit 25:613–632

Collier P (2016) Das Etikett Flüchtling geht am Problem vorbei. http://www.ipg-journal.de/interviews/artikel/das-etikett-fluechtling-geht-am-problem-vorbei-1717/. Accessed 13 Mar 2019

da Conceição-Heldt E, Koch M, Liese A (2014) Internationale Organisationen als Forschungsgegenstand. Oder: "Über Blinde und die Gestalt des Elefanten". In: Ed C-H, Koch M, Liese A (eds) Internationale Organisationen: Autonomie, Politisierung, interorganisationale Beziehungen und Wandel. Nomos, Baden-Baden, pp 4–27

Deschamp B, Azorbo M, Lohse S (2010) Earth, wind and fire: a review of UNHCR's role in recent natural disasters. UNHCR Policy Development and Evaluation Service, Geneva

Dingwerth K, Kerwer D, Nölke A (eds) (2009) Die Organisierte Welt: Internationale Beziehungen und Organisationsforschung, 1. Aufl. Internationale Beziehungen, vol 12. Nomos, Baden-Baden

Dingwerth K, Lehmann I, Reichel E, Weise T, Witt A (2014) Many pipers, many tunes? Die Legitimationskommunikation internationaler Organisationen in komplexen Umwelten. In: Ed C-H, Koch M, Liese A (eds) Internationale Organisationen: Autonomie, Politisierung, interorganisationale Beziehungen und Wandel. Nomos, Baden-Baden, pp 186–212

Ege J, Bauer MW (2013) International bureaucracies from a public administration and international relations perspective. In: Reinalda B (ed) Routledge handbook of international organizations. Routledge, New York, pp 135–148

Enkler J, Grohs S (2017) An den Grenzen der Lernfähigkeit? Lernkurven internationaler Verwaltungen nach der Finanzmarktkrise. dms-der moderne staat-Zeitschrift für Public Policy, Recht und Management 10:251–270

Franck M (2015) When leaving your home is the last option for survival: climate change and human mobility, presentation. COST Action 1101 Climate Change and Migration; Knowledge, law and policy, and theory, Durham, UK

Gemenne F (2015) Disasters and displacement in a changing climate: one good reason to speak of 'climate refugees'. Forced Migr Rev 49:70–71

Glover T, Russell S (2001) Overseeing the refugee convention: coordination with UNHCR and states. ICVA Working Paper No. 7. International Council of Voluntary Agencies, Geneva. https://papers.ssrn.com/sol3/papers.cfm?abstract_id=294181. Accessed 2 July 2019

Guterres A (2012a) Foreword. In: Betts A, Loescher G, Milner J (eds) UNHCR: the politics and practice of refugee protection, 2nd edn. Routledge, London, pp xx–xxi

Guterres A (2012b) Opening statement. In: Pledges 2011: Ministerial intergovernmental event on refugees and stateless persons, Geneva, 7–8 Dec 2011, Geneva

Haas EB (1990) When knowledge is power: three models of change in international organizations. University of California Press, Cambridge, MA

Hall N (2013) Moving beyond its mandate? UNHCR and climate change displacement. J Int Org Stud 4:91–108

Hall N (2015) Money or mandate? Why international organizations engage with the climate change regime. Glob Environ Polit 15:79–97

Hall N (2016) Displacement, development and climate change: international organizations moving beyond their mandates. Routledge, Abingdon

Hall N, Woods N, Kabra SS, Taranova Y, Kellermann M, Batten H (2015) Effective leadership in international organizations. World Economic Forum

Hammerstad A (2014) The rise and decline of a global security actor: UNHCR, refugee protection, and security. Oxford University Press, Oxford

Janmyr M (2016) Advancing UNHCR accountability through the law of international responsibility. In: Sandvik KB, Jacobsen KL (eds) UNHCR and the struggle for accountability: technology, law and results-based management. Routledge, New York, pp 46–63

Kälin W (2010) Conceptualizing climate-induced displacement. In: McAdam J (ed) Climate change and displacement. Hart, Oxford, pp 81–104

Kita SM, Raleigh C (2018) Environmental migration and international political security: rhetoric, reality and questions. In: McLeman R, Gemenne F (eds) Routledge handbook of environmental displacement and migration. Routledge, New York, pp 356–370

Loescher G (2014) UNHCR and forced migration. In: Fiddian-Qasmiyeh E, Loescher G, Long K, Sigona N (eds) Refugee and forced migration studies. Oxford University Press, Oxford, pp 215–226

Loescher G, Milner J (2011) UNHCR and the global governance of refugees. In: Betts A (ed) Global migration governance. Oxford University Press, New York, pp 189–209

Loescher G, Betts A, Milner J (2012) The UNHCR: the politics and practice of refugee protection, 2nd edn. Routledge, New York

McAdam J (2010) Introduction. In: McAdam J (ed) Climate change and displacement. Hart, Oxford, pp 1–8

McAdam J (2011) Environmental migration. In: Betts A (ed) Global migration governance. Oxford University Press, New York, pp 153–188

McAdam J (2014) Creating new norms on climate change, natural disasters and displacement: international developments 2010–2013. Refuge 29:11–26

McAdam J (2016) How do we deal with the prospect of increased climate migration? UNSW Law Publication. http://www.law.unsw.edu.au/news/2016/12/how-do-we-deal-prospect-increased-climate-migration. Accessed 13 Mar 2019

McAdam J (2017) From the Nansen initiative to the platform on disaster displacement: shaping international approaches to climate change, disasters and displacement. UNSW Law J 39:1518–1546

Methmann C (2014) Visualizing climate-refugees: race, vulnerability, and resilience in global liberal politics. Int Polit Sociol 8:416–435

Nansen Initiative (2011) The Nansen Conference: climate change and displacement in the 21st century. Norwegian Refugee Council/Internal Displacement Monitoring Centre, Oslo. https://www.refworld.org/docid/521485ef4.html. Accessed 2 July 2019

Nielson DL, Tierney MJ, Weaver CE (2006) Bridging the rationalist–constructivist divide: re-engineering the culture of the World Bank. J Int Relat Dev 9:107–139. https://doi.org/10.1057/palgrave.jird.1800084

Oels A (2012) From "securitization" of climate change to "climatization" of the security field: comparing three theoretical perspectives. In: Scheffran J, Brzoska M, Brauch HG, Link PM, Schilling J (eds) Climate change, human security and violent conflict: challenges for societal stability. Springer, Berlin, pp 185–205

Oels A, von Lucke F (2015) Gescheiterte Versicherheitlichung oder Sicherheit im Wandel: Hilft uns die Kopenhagener Schule beim Thema Klimawandel? Zeitschrift für Internationale Beziehungen 22:43–70

Pécoud A (2015) Depoliticising migration: global governance and international migration narratives. Palgrave Macmillan, London
Rothe D (2016) Securitizing global warming: a climate of complexity. Routledge, New York
Scheel S, Ratfisch P (2014) Refugee protection meets migration management: UNHCR as a global police of populations. J Ethn Migr Stud 40:924–941
Siebenhüner B, Biermann F (2009) International organizations in global environmental governance: epilogue. In: Biermann F, Siebenhüner B, Schreyögg A (eds) International organizations in global environmental governance. Routledge, New York, pp 264–269
Steffek J, Kortendiek N (2018) Participatory governance in international organizations. In: Heinelt H (ed) Handbook on participatory governance. Edward Elgar, Cheltenham, UK, pp 203–224
Trombetta MJ (2014) Linking climate-induced migration and security within the EU: insights from the securitization debate. Crit Stud Secur 2:131–147. https://doi.org/10.1080/21624887.2014.923699
Türk V, Eyster E (2010) Strengthening accountability in UNHCR. Int J Refugee Law 22:159–172

FSC
www.fsc.org
MIX
Papier | Fördert
gute Waldnutzung
FSC® C083411

Kolibri 360 GmbH
Ferdinand-Jühlke-Straße 7
99095 Erfurt, Deutschland
produktsicherheit@kolibri360.de